Darwin Meets the Buddha
Human Nature, Buddha Nature, Wild Nature

*All things may be impermanent,
but some things are more
permanent than others.*

Paul A. Keddy

DARWIN MEETS THE BUDDHA
Human Nature, Buddha Nature, Wild Nature
Paul A. Keddy

Text © Paul A. Keddy, 2020
All rights reserved

Editing & design: John Negru
Cover photo: Paper Street Design / Shutterstock
Author photo: Cathy Keddy

Published by
The Sumeru Press Inc.
Ottawa, ON
Canada

LIBRARY AND ARCHIVES CANADA CATALOGUING IN PUBLICATION

Title: Darwin meets the Buddha : human nature, Buddha nature, wild nature / Paul A. Keddy.
Names: Keddy, Paul A., 1953- author.
Description: Includes bibliographical references and index.
Identifiers: Canadiana 20190223405 | ISBN 9781896559575 (softcover)
Subjects: LCSH: Evolution (Biology)—Religious aspects—Buddhism. | LCSH: Evolution—Religious aspects—
   Buddhism.

Classification: LCC BQ4570.E85 K43 2020 | DDC 294.3/36576—dc23

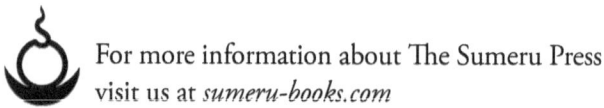

For more information about The Sumeru Press
visit us at *sumeru-books.com*

Contents

Introduction 5
The Men and Their Theories 15

1. Craving for Resources: Desire, Dissatisfaction, and Suffering . 25
2. Living an Illusion: Mind as Cocoon 39
3. The Primate Prison: The Origin of Self. 51
4. Selective Memory: Maintaining the Illusion 65
5. The Urge to Impress: Priests, Kings, and Dominance Hierarchies 77
6. Killing Minds and Killing Fields: Interference, Competition, and Aggression. 99
7. Insatiable Consumption: When Big Brains Meet Big Animals . 115
8. Getting Along: An Ecological View of Compassion 133
9. A Brief History of Life: Co-operation and Community 145
10. Meditation in Action: Seeing Through the Simulation. 165
11. Enlightened Society: The Evolutionary Imperative. 191

Acknowledgments 211
Appendix: How to Start Now 215
Further Reading 217
Figure Credits 233
Index. 235

Introduction

DID YOU HEAR? Charles Darwin has invited Siddhartha Gautama to hike the Appalachian Trail through the Great Smoky Mountains. How would you like to join them and eavesdrop on the conversation—just about any topic is imaginable. Maybe you will even have the opportunity to ask a question or two. I have reserved a place for you on this expedition. Of course, you will have to pay attention as well as carry your own food and your own tent, and you may even learn something about forests and camping as we follow the conversation.

Yes, Darwin and the Buddha, on one trail. For some reason, people seem determined to draw a firm line between secular and spiritual life. Over the years I have enjoyed many expositions and books on the teachings of Darwin, and it is clear that his work has a great deal to say about why humans behave the way they do. I have also enjoyed many expositions on the teachings of the Buddha, and it is apparent that the practical instructions of his Eightfold Path also have a great deal to say about why humans behave the way they do. Indeed, both world views have more in common than you might first suppose. This is because both are grounded in describing what *is*. Not what we wish were true, but what *is*. So, both men have a lot to talk about when it comes to the human condition.

Overall, I have found that most Buddhist writers and visiting Buddhist teachers know only half the story. Indeed, attempts to bring science into Buddhism usually mention not biology but physics, and the 'physics' is often just painfully inaccurate or an irrelevant use of words, such as relativity or quantum mechanics. We are, however, biological beings, and in this conversation I wanted to focus on human biology and behaviour and how this relates to meditation practice altogether. Overall, this book began with the intention of assisting Buddhists who need some knowledge of how their meditation practices are rooted in human evolution and how they should rouse themselves to be more useful in protecting wild nature. The selection of topics was, therefore, somewhat directed by my personal experience of areas where biology and history have something useful to say to one another.

As the book progressed, it struck me that it might also work the other way around, rather like a reversible jacket. That is, the book might equally be useful to secular readers who are curious about whether Buddhism has anything useful to say about our lives as modern human beings. Where else, I wondered, could readers obtain an introduction to Buddhist practices that is grounded in biological reality rather than mythology? Or a book that not only talks about compassion for all living beings but dares to include those non-human beings living in rainforests and on coral reefs? There was not a single book in my library that could fill this role. This book does it.

During the project, it further occurred to me that this book might also be helpful to an emerging third audience—younger people who are struggling to understand their lives and their place in society, and who know rather little about either Darwin or the Buddha. Knowing a little more about both these extraordinary men will actually assist each of us to lead a life that has less suffering and more sanity. Such a view of life may also help us understand why certain problematic behaviours keep emerging, both in ourselves and in others. It may even provide some guidance on how to work with problems more skilfully. Life is not easy, and both Darwin and the Buddha have something to say about why we find this to be the case. They were both keen observers of the human condition. Darwin was perhaps more interested in how people came to be in the first place, particularly in their exterior form (although he did write a good deal on human emotions). The Buddha, living earlier, was mostly interested in what people experienced, and hence tended to focus more on what we might call the interior of their lives. So, this book is perched on the edge—looking outward into wild nature and inward into human nature. The book, therefore, offers a kind of binocular vision. Since binocular vision is common in the animal kingdom, it apparently has certain merits. And we know that binoculars are very helpful when you are making acquaintance with wild nature.

The manuscript has been a long-term project and a difficult one. I am now going to tell you a little bit more about how it came to be and my qualifications to write such a book. My guide for this introduction is George Bernard Shaw, who wrote many a long preface to his books and plays. If you don't like a long opening, then just jump to Chapter 1. There is also a short section on the basic teachings of Darwin and those of the Buddha which you may find helpful as a quick review prior to leaping into the first chapter.

Let us return to the topic of audience. The first challenge has been a practical one: how do you write one book for some quite different audiences? The point is to bring people together, not to create dissention. I continually wrestled with this challenge.

First let us consider the original audience for this manuscript, fellow Buddhists. There are Buddhists with whom I have practiced for decades who seem fearful that talking about Darwin is an attempt to explain away their spiritual life. Some of them seem to despise modern science and, instead, seem to be attempting to resurrect a medieval view of reality. An example. Some years ago, at a Buddhist retreat for families in Vermont, I was invited by some parents to lead a 'nature walk' for the children so they could be introduced to the surrounding forest. Just as I was getting under way, a senior staff person hustled out of the main building and told me in front of the children that I would have to stop immediately for I was undoing her work with the children's minds. (Yes, it surprised me so much that I really did surrender to that authority figure and let her take the children back into the building.) If only that incident were an exception. Instead, I keep meeting practitioners who seem to prefer emoting and virtue signalling to the use of evidence-based thinking. Such people are actually causing themselves (and others) harm. The current human situation absolutely requires each of us to have at least a passing familiarity with topics including evolutionary biology, the biological basis of human hierarchies, the fundamental differences between eggs and sperm, the importance of X and Y chromosomes, and the urgent need to protect wild nature. These and other biological topics have a profound impact on how we view ourselves and how we treat one another as well. Let me repeat this point: some familiarity with certain aspects of basic biology will help Buddhist practitioners with their own lives, and it will simultaneously help them to be better neighbours for other living beings. The Bodhisattva vow, after all, is a commitment to care for all living beings, which includes whales and rhinoceros and sea turtles, not to mention wilderness areas altogether. Is it possible that some or even many people are misunderstanding the Buddhist path and using it as an excuse to withdraw from reality and avoid caring for wild places and wild creatures? That is currently an open question.

Now for the scientists. There are biologists I have worked with for decades who probably wonder why a serious scholar with a reputation for empiricism would be involved with Buddhism at all. Here, for them, is part of the explanation. We are living at a time when humans desperately need solid, basic science to guide conservation and the wise use of the Earth, yet basic science is seriously distorted by greed and egotism. I have actually written scientific papers about the degree to which basic science in ecology is being distorted by human behaviour, and it was these concerns that inspired me to look more into Buddhist meditation in the first place. It offers a kind of experiential window into the inner workings of the human mind. Of course we need good science to solve (certain) environmental problems, but throwing money at problems may actually increase the distortion inherent in those problems. An example. I worked in Louisiana on coastal wetlands restoration for eight years and would have to say that greed, self-delusion and egotism were huge problems. In fact, you could say that a system ostensibly set up to restore coastal wetlands had, instead, become corrupted into a program that allowed professors to supplement their retirement funds and universities to bleed off federal money to construct new buildings and pay their administrators. The fate of the wetlands themselves and the wild creatures in them had become quite irrelevant. Mardi Gras and plastic beads were a far higher priority. The coastal restoration system, then, had become distorted by what the Buddha called 'the three poisons': greed (for federal money), ignorance (of good science) and aggression (toward anyone who pointed out the flaws in the system). The restoration venture had thus become a vivid illustration of what the Buddha called *samsara*. When I was a young scientist, I genuinely believed that if we understood the science behind environmental problems, conservation would naturally follow. Forty years later, my life experience intrudes and says that it might not be so. It may indeed be the case that effective responses to obvious problems are being delayed by human behaviours that are deeply rooted in the past. Some of these behaviours were diagnosed by Gautama Buddha two and half millennia ago. How might our success at conservation improve if we had a better familiarity with human emotive states? I use the word familiarity on purpose: it is one thing for us to know about recent advances in psychology, some of which I mention here, but it is quite another to feel them acting in our own minds. It is like the difference between seeing an alligator in a book and meeting one in person while wading chest-deep in a wetland.

At this point you may ask, how is it possible that one can write for two such apparently-different audiences? Perhaps it is a foolish errand and will just annoy both. What has kept me going on this project is the conviction that beneath the differences there are real common interests: how do we accept our limitations, how do we live as decent people, how do we organize compassionate societies, and how do we care for all the other sentient beings with whom we share the planet? Modern terms like intersectionality and our fascination with diversity are mostly just excuses for us to ignore our common interests as human beings. Speaking of caring for all sentient beings, this does not just mean your friends, relatives and pets, but rhinos, elephants, sea turtles, salamanders and tree frogs, and even more humble creatures. I am surprised how little attention they receive from my fellow Buddhists. The whole Buddhist path, says Bhikkhu Bodhi, is anchored in having *Right View*. Part of that right view surely includes a view of life expansive enough to include the biological nature that is inside us, and the wild nature that is outside us.

Overall, it seems as if biology and spirituality still each have their own communities and there is little meaningful exchange. If anything, the gulf may have actually widened over the past decade. My secular friends and fellow biologists mostly seem to avoid 'religion' as a topic of conversation, probably for fear of causing offense or perhaps from simple lack of interest. Yet many of these same friends have a deeply spiritual relationship with wild nature, which is no small matter. So they take long canoe trips or hike in the mountains or go camping. Such trips might be enriched by some appreciation of how our own mind conditions our experience and how it is that we find the experience desirable. We don't even need to bring up the name of the Buddha. And, further, many spiritual leaders have, themselves, spent time alone in the wilderness, including Jesus and St. Francis of Assisi. Many of my secular friends have spent more time in the wilderness than my Buddhist friends, and I think my Buddhist friends have missed something important. But wilderness trips themselves can easily be distorted by the human ego, as when they turn into athletic endurance contests aimed at covering as many miles as possible in a day or opportunities to display high-tech camping gadgets. The same is true of other activities. Even academic science and conservation biology might, as noted, have better outcomes if we were less fixated on our sense of self-importance, the size of our grants, and our desire to climb in the academic hierarchy. Overall, then, self-examination is likely to be useful in even the most secular life.

And now for the third audience, younger people. Although I have lived in the forest for a decade, I do see younger people as visitors and at various events. I am continually struck by how little they seem to know about history, biology, or Buddhism. The internet and social media seem to be drowning out important information in a sea of ephemeral thoughts and emotions, each up-voted or down-voted. Hence, the social media seem, by and large, to be increasing the amount of human confusion. Indeed, if we take the view that the Buddhist path opens us up to experience our own confusion, it is just possible from this perspective that the internet has become a kind of amplifier, creating levels and styles of confusion that the Buddha himself could not have imagined. Apparently many young people also have little connection with wild nature. I come to this conclusion from hundreds of causal conversations where I have paid careful attention to topics that arise. Few incidents of real wild nature enter into conversation naturally. Sometimes I deliberately raise the topic by asking something apparently innocent. Questions like, "What was the first bird you saw this spring?" or "Which park did you last hike in?" are met all too often with a strange look as if I had asked about their recent descent into a black hole. Here is the awful conclusion that I have been forced to accept: it is not just that they have nothing to say, but that they do not even understand why I am asking. Of course, I am testing whether they are connected to reality. Each interaction with wild nature connects our mind to what *is*. Overall, I tend to think that we are now seeing one of the opportunity costs of the internet: every hour online is an hour not spent interacting with real birds or real plants or real forests. It is even possible that some kinds of meditation actually harm such people, by further detaching them from reality. In some essential way, reality itself is a part of the path and wild nature is therefore necessary for sanity.

My purpose here, for all these audiences, is to present some basic Buddhist teachings on the human condition and some basic human biology and history, then explore why humans are the way they are and why we experience the world in a confused way, and finally look at how this affects the way we live our lives and what this might have to do with the future of all the other species with whom we share this planet. This inquiry may not make everyone happy. Science is not the search for happiness or pleasure, but the search for truth. The same might be said for the teachings of the Buddha. To pick a third authority figure, even Jesus said that the truth would set us free. He did not say that the truth would bring us pleasure or find us a soul mate or pay off our mortgage.

Finally, a note on style. For most of my life I have had to write heavily-referenced scholarly papers and books, the joyful exception being when I write natural history articles for newsletters and newspapers. With *Darwin Meets the Buddha*, I wanted to avoid that somewhat ponderous style and make it a very readable book. So, there are few explicit references. There is a list of source books and papers for each chapter at the end of the book. Yes, I own nearly all of them.

I should probably also say something now about my qualifications for this project. At least that is what my publisher suggested. That seems reasonable, as we are going to cover a lot of ground, and to some extent you are going to have to trust my understanding of both biology and Buddhism. I have taken the task of this book upon myself for the simple reason that no one else has done so to my satisfaction. The current situation is quite harmful to Buddhism and possibly to biology too.

Let us start with the biology. I have been a biology student as long as I can remember, having always been attuned to seeing turtles, salamanders and frogs in the landscape around me. When I was about ten years old, my parents were surprised that I could spot a turtle on a log as they were driving along a prairie highway at 50 miles per hour. I left home early, at the age of 16, to study science at York University in Toronto and then Dalhousie University in Halifax. During the summers when I was an undergraduate, I worked as a park naturalist in Algonquin Provincial Park. (This job, of explaining the biology of the park to visitors, is sometimes called a park interpreter or a park ranger.) The point is that all year round my time was spent studying science in general and biology in particular. I then became a professor of biology. Over my professional career I have so far worked at three universities—each with quite different academic cultures, taught thousands of students, published over 100 scientific papers, studied many different forests and wetlands, and written multiple books. All this activity has been focussed on one problem: how do we build our understanding of wild nature as effectively as possible, and how do we apply this knowledge to protect wild species and wild places? Along the way, there have been some scientific prizes, including a lifetime achievement award from the Society of Wetland Scientists. And, for those who like to count, I was designated a highly-cited researcher by the Institute for Scientific Information. I'm still working actively as a scientist. You can read more on my web site (drpaulkeddy.com).

And now for my Buddhist credentials? Yes, my friends, even Buddhists have their own set of criteria for judging one another. They may be even worse snobs than some academics for enquiring how you were trained and what you have studied! So here is my Buddhist background. I began studying meditation in the late 1970s when a friend from British Columbia first introduced me to Buddhist meditation and the Buddhist path. It was a very different time, then, as the mindfulness movement had yet to permeate North America, and access to the Buddhist teachings was much more limited. It was therefore a surprise, of sorts, to find out that several millennia ago the Buddha had described the habitual patterns of human beings and documented the confusion and suffering that results. Siddhartha Gautama, the man we now call the Buddha, had even coined a word for the whole situation, samsara. It all comes down to three causes, said the Buddha: passion, aggression, and ignorance. And by ignorance he meant not lack of information, but a deliberate wilful intent to evade reality, a term we now call wilful blindness. And, he went further to suggest there were tools we could use to escape the grip of samsara upon our lives. So, gradually the Buddhist Eightfold Path became woven into my life. In Ottawa I found several qualified young teachers who offered weekly classes and meditation nights in my own community. With time, I ventured forth into other opportunities, including a month-long group retreat called a *dathun*, week-long solitary retreats in the forest, and other group meditation retreats high in the mountains. Three senior teachers stand out in my memory. Khenpo Sonam Rinpoche taught the foundations of Mahamudra in Ottawa, and I later worked with him on a small publication that was called *Enlightened Attitude*. Around this time, Chögyam Trungpa Rinpoche, another Tibetan Buddhist, was presenting a secular path based upon Buddhist foundations, called the Sacred Path of Warriorship. He clearly stated that one did not have to become a monk, or a Tibetan, or a conventional 'Buddhist' of any sort; rather, spiritual practices could be woven into the life of an ordinary householder. Eventually I took formal refuge vows with a third Tibetan teacher, the late Dilgo Khyentse Rinpoche, in a ceremony in which I committed myself to the Buddhist path. During these events and over several decades, I was also provided with some exposure to Zen and Theravada Buddhism, each of which has rather different cultural norms and styles of teaching from those imported by Tibetans. For those of you who are curious about the milieu at the time, the book *How the Swans Came to the Lake*, by Rick Fields, surveys how many kinds of Buddhist teachings were spilling into North America, particularly after the Second World War. The point is that while I have had some training, the Buddhist material in this book comes not from credentials of being some sort of professor of religious studies nor some sort of officially designated 'lineage holder' but simply being a human who has studied, practiced, and made a genuine effort to live this path while remaining engaged in his own profession. Along the way, over nearly four decades, my intention has always been to select Buddhist practices that are useful for modern people. The Buddha lived several thousand years ago, after all, and would certainly teach differently now than he did then. To exactly follow his example, abandon our homes and our families, and start walking from one city to another, is probably not what he would recommend today. In *Buddhism Without Beliefs*, Stephen Batchelor ventures the opinion that the Buddha would possibly not include either karma or rebirth if he were teaching now. Our modern challenge is to include what we now understand about the human psyche and apply it to how we live and how we behave, with particular emphasis upon how we treat other humans and non-humans. Moreover, in the Buddha's day it was relatively easy to find wild places for retreat; it is no longer so, and a modern Buddhist path now has to include caring for the diminishing wild places of the Earth. I will say more about this later. Speaking of forests and wild places, for the last ten years I have been living at the very end of a dead-end road, surrounded by a square mile of protected forest and wetland.

If this all seems like more, far more, than you wanted to know about me, I can assure you that other readers will feel differently, and they will likely want to know even more. Biologists have a set of criteria for judging other human beings, such as what journals they publish in, how often their work is cited, and how much grant money they raised for research. Buddhists have another set of criteria, such as the names of the teachers with whom you have studied, 'programs' you have been completed, and the empowerments that you have collected along the way. My intention here was to document that I am somewhat qualified for the task at hand. I can give you some sense of what we have learned from Charles Darwin and Siddhartha Gautama and how it might apply to your own life. After that, it is entirely up to you.

In conclusion, this is a book with a general approach to a broad problem. Many of the examples are rarely used in Buddhist teaching which is exactly why I chose them. I freely admit to having only a limited knowledge of the French Revolution, the Second World War, the Bolshevik Revolution, animal behaviour, and human neurophysiology. In spite of my limitations, casual conversations reveal that often I know more about these topics than many people and certainly more than most younger visitors. Someone has to make the effort to put all this together, even at the risk of offending the specialist. No doubt there are errors and oversights in my understanding, but for some time I have seen a need to try to create a larger picture.

It is time to get back to our premise. What would happen if Charles Darwin and the Siddhartha Gautama had met and struck up a relationship? What might they have to say about human nature, wild nature, and the events unfolding around us today?

The Men and Their Theories

THE TITLE ERRONEOUSLY SUGGESTS that Darwin met the Buddha when, in fact, the Buddha lived and died in India some 2,500 years ago and Darwin lived and died between 1809 and 1882 in Victorian England. Although they did not meet personally, their legacies of ideas have lived on. Before discussing their theories, let me introduce the men themselves. In fact, each of these men has spawned a virtual industry. Darwin left enough notes, letters, journals, and books to support a small army of scholars. The Buddha's oral instructions were first memorized and then eventually written down in Pali, and these writings and his oral instructions have also spawned innumerable books by secular and religious scholars.

For biographical details on the Buddha, I have relied mostly on *The Historical Buddha* by H.W. Schumann, while for Darwin I have used *Darwin* by Adrian Desmond and James Moore. Both of these books are filled with rich details about their personal experiences. Consulting *Chambers Biographical Dictionary* reveals a few succinct details:

> Prince Gautama Siddhartha (Buddha) (c. 563 - c. 483 BCE) was born into the Sakya tribe in Nepal. At about the age of thirty he left the luxuries of court life, his wife, and his earthly ambitions for the life of an ascetic; after six years of self-denial, he turned to a more gentle contemplative life, and according to tradition, attained enlightenment near Bodh Gaya in Bihar, taught for forty years and died near the age of 80.
>
> Charles Robert Darwin (1809-1882) was born in Shrewsbury England, and studied medicine at

Edinburgh. He went to sea on the Beagle with captain Fitzroy, in spite of continuous ill health published numerous books on topics including coral reefs, earthworms, barnacles, and orchids, and in 1859 published *The Origin of Species by Means of Natural Selection.*

A human leads a life that we can try to evaluate in moral terms as being either good or bad. A theory is different. A theory is good or bad only to the degree that it accurately describes reality; that is, it bears some similarity to what we can observe. When scientists talk about a good theory or a bad theory, they are not making a moral judgment, they are talking about goodness-of-fit. The better the model fits the world, the better a theory it is. Of course, humans also find it easy and natural to make moral or ethical judgments, and if a theory appears to have bad consequences, people might easily describe the theory as bad because they do not like those consequences. Yet, it really is nonsensical to ask whether gravity is good or bad or whether the speed of light is good or bad. It is. Period. End of discussion. People may die because trees fall upon them or because they fall off cliffs, but it would be nonsense to label Newton as evil for having discovered gravity. In a similar way, the Buddha and Darwin described certain aspects of reality. The moral consequences of their theories may disturb us, but if the theory works, then it is a good theory.

Since the natural world does not always meet with our expectations, many 'good' theories (that is, theories with convincing goodness-of-fit to our experience) have an annoying tendency to demolish our cozy and long-held views. For example, in spite of so many claims by so many religious teachers, we are not at the centre of creation. Of course it feeds our sense of self-importance to imagine that we are, but careful observation and calculation show that we are orbiting a minor sun, one of only 200 billion stars in the Milky Way. It would therefore seem, at this larger scale, that humans are less important than we would wish to be. That we are orbiting but one of hundreds of billions of suns at the edge of but one of billions of galaxies seems to describe accurately our state of affairs. We may not like it, but knowing it to be the case, it is up to us as humans to decide what, if any, ethical, moral, or spiritual consequences follow from the observations.

This book argues that both the Buddha and Darwin have left us similar 'good' theories. They are good theories because they describe the state of the world accurately. Indeed, the theories work so well that they could be (and indeed have been) accused of simply describing completely self-evident aspects of reality. If, however, their work were so self-evident, brilliant minds would not have been required to uncover it. Perhaps the very best kind of theory is one that seems, with perfect hindsight, to be self-evident—the goodness-of-fit between the theory and the observations is so nearly perfect that surely it would be obvious to anyone. Well, if it were so obvious, then why did it take so long for humans to discover it? This latter question is one that the book will address, at least tangentially.

For those unfamiliar with the work of the Buddha or the work of Darwin, I provide here a brief summary, given in four propositions or observations, about how the world works.

Darwin's theory

1. Production of young

Every organism on earth has the potential to produce vast numbers of young. Although humans reproduce rather more slowly than microbes or rodents, men at least can produce billions of sexual gametes in one lifetime, and a man could produce thousands of children. The average elm tree or mushroom makes humans look like slackers; an elm tree can produce a million offspring in a hundred-year lifetime; a mushroom could easily produce ten billion. But whether we have a mushroom or a human, only one of the offspring needs to survive for that individual to replace itself. Nature, then, is very wasteful; the other 9,999,999,999 spores from the fungus, the other 999,999 seeds from the elm, and the other 999,999,999 sperm cells of a man will all die. The individual is utterly expendable.

2. Variation

There are complex biological mechanisms hidden within the process we call sexual reproduction, mechanisms such as meiosis, to insure one thing: that nearly every sperm and egg is different from the next. Thus, every time a sperm and egg unite, a completely new and fresh genetic combination is formed. Every seed of the elm, every spore of the mushroom, and every sperm from a man carries a mixture of genes for a completely new biological entity. Every new generation contains the raw material for a striking array of new and different life forms. The variation generated by the mixing of genes during reproduction is not sufficiently appreciated, but it is the major source of variation in life on Earth. Superimposed upon this variation generated by sexual processes are occasional random events of damage to genetic material. The results of such damage are called mutations, and they can create entirely new genetic sequences to be mixed into the pool of sexual variation. Many people misunderstand and think that mutation is the main cause of variation in life when, in fact, it is sexual recombination. From the lowly mushroom to the proud rhinoceros, there are countless internal safeguards to ensure that offspring are slightly different from their parents.

3. Selection

Of the million elm seeds produced, only one need survive. Should the climate become dry, it is probable that many of the 999,999 seeds and seedlings that die will be the ones most sensitive to drought. The causes of death may vary with the genetic composition of the seedling. Perhaps their leaves have an inadequate wax coating on the epidermis, perhaps the roots of a seedling grow too slowly to reach the water table, or perhaps the bark is too permeable to water. Drought may not be the selective force; equally, there might be heavy grazing by deer. In this case it is likely that many of the 999,999 that die will be those most sensitive to having their leaves or growing tips eaten. Perhaps the leaves lack sufficient numbers of hairs, perhaps the foliage lacks foul-tasting chemicals, or perhaps the seedlings are slow to recover from damage. Neither drought nor deer may be the problem. If instead there is recurring fire, many of the 999,999 that die will be the ones with seeds that lack thick walls or seedlings with thin bark or a cambium that is easily killed by heat. More examples are possible. The point is that the environment constantly challenges life forms and eliminates (selects against) those that are not adapted to existing conditions. That is, it filters out the variants that cannot survive the present environment and leaves behind those variants that have survived existing conditions. I repeat the word 'existing' conditions because natural selection does not plan ahead—there is no cosmic mind worrying or strategizing about whether next year might be wet or whether next year will have more deer or whether next year will have more fires. There is nothing like this at all. There is only an entirely passive process whereby the environment destroys those offspring that tend to be the least suited to current conditions. There are a lot to be destroyed, and those very few that do survive tend to be those best adapted to the current conditions.

Whether they have more leaf wax or fewer stomata to survive drought; bad tasting leaves or toxic hairs to discourage deer; thick bark or resistant seeds to tolerate fire, whatever happens to work tends to survive. Since an organism cannot predict what the future will be like, there is an advantage to producing variable offspring so that at least a few out of the millions of elm offspring will have the appropriate combination of genes to tolerate the unpredictable future environment. Humans consciously practice a similar strategy when they diversify their investments among a number of companies since they cannot predict the future and do not know which company will be the one to succeed.

One commonly hears the view that evolution cannot work because 'mere chance' cannot produce life. The diversity of life forms and their close fit to their environments cannot, the argument goes, have arisen by 'mere chance' mutation (and therefore, the line of argument usually continues, evolution cannot occur). The chance event of mutation, as I have indicated above, is nearly irrelevant to the process of natural selection. The 'chance' element in the process is actually generated largely by sexual mixing of pre-existing genes. The order is created not by 'mere chance' but by selection.

Selection is a consequence of the mismatch between the re-shuffled genetic materials and the environment to which the organism is exposed. When only one out of a million offspring can survive, there is a great deal of raw material for selection to act upon. The extremely close fit between organisms and their habitats is, therefore, not a matter of chance at all. Selection occurs because the environment has served as a template to which all organisms are forced to conform or die. When Darwin talked about survival of the fittest, this is what he meant. The fittest are those whose genetic predisposition provides the best short-term fit to the environment—the closest fit to the habitat template. This is an entirely passive process. If an individual doesn't fit, it dies. If it does, it lives and gains the opportunity to reproduce.

4. Inheritance

Although sexuality ensures that offspring are not exactly like their parents, inheritance does occur. Offspring tend to be somewhat like their parents, yet also different. This is why relatives have so much fun trying to discover which traits of parents turn up in children and grandchildren. To study the process of selection, biologists often focus on the dissimilarity of parents and their young; this, after all, provides the evidence of natural selection at work. But offspring and parent similarity, however, is the final factor that allows evolution to occur. The elm seed that does live to become a tree has many of the genes of its parent trees, and so these genes continue to persist and tend to increase in elm trees relative to other genes. Genes that are not suited to current conditions have costs and tend to be lost over time. For example, if deer grazing is heavy, the genes that make leaves taste good will gradually be lost, resulting in a relative increase of those that make leaves taste bad; if there is drought, the genes that put many stomata in leaves will decline in frequency relative to those that produce waxy leaves. All offspring carry some (almost exactly half) of the parents' genes, and the flow of these from generation to generation is what allows the chain of life to continue through millennia. Somewhere buried within each of us are genes that have passed through bacteria, amphibians, reptiles, mammals, and our past generations of humans to find their expression in us today.

A twentieth century student of Darwin, Richard Dawkins, has explained that we may better understand the process of evolution by considering it at the scale of genes rather than organisms. Organisms, he argues, are just a gene's way of producing more copies of itself. Each organism is a gene factory, and the genes that survive and multiply do so because they have made gene factories that are likely to persist and efficiently construct many more copies of those genes that built them. Survival of the fittest, then, really means survival of the factories that are most successful, factories that are busy producing copies of the genes that serve as their blueprints. From the perspective of the gene, the organism itself is relatively unimportant so long as it produces more copies of the genes.

Genes shuffle and they mix with genes from other factories. The resulting mixtures start to build new gene factories. The environment destroys all but a few factories. The remaining factories make many more copies of the genes in their blueprints. This process has taken the Earth from one-celled prokaryotic life to complex multicellular organisms that are conscious of themselves. It is *not* a random or chance process; rather it is a process that is inevitable, given the way the environment ruthlessly eradicates any creature that is unsuited to conditions.

In summary, the processes that have produced life are entirely passive reactions of genes and environments, no more of a mystery, and no more or less moral than the nuclear reactions at the heart of giant stars, the orbital exchanges that occur in chemical reactions, or the movement of planets around the sun. That is the way it *IS*. It matters not whether humans find it good or bad. It matters not what we think the moral consequences are because nature is amoral. Evolution is neither good nor bad, it is just an accurate description of our common origins and how life continues to unfold.

Buddha's theory

1. Life is suffering

The first statement of the Buddha's discovery is that all life experiences suffering. The exact word was *dukkha*. Sometimes it is best just to use dukkha, which avoids the challenges of translation and misunderstandings about the word 'suffering'. The truth that life is suffering is known as the First Noble Truth. There are the gross kinds of suffering, such as the pain of birth, sickness, old age, and death. And then there are the subtle kinds: fear, anxiety, misery, and depression. Some of the former kinds of suffering have been successfully alleviated by technology: improvements in agriculture reduce famines, clean water reduces disease transmission, medicines can dull physical pain, doctors can heal broken bones. There have been advances here that the Buddha could never have imagined. But it seems that as fast as humans heal these old problems, new ones arise to replace them. Not only are there new diseases and new problems due to a limited environment with growing populations, but the subtle pain of confusion and anxiety continues to haunt humans around the world.

Some have suggested that the word 'suffering' is a mistranslation of the Buddha's actual teaching. The word 'dissatisfaction' may come closer to what he meant. Life is full of dissatisfaction because nothing that humans can do is ever enough to provide more than a momentary escape from primordial pain and confusion. In spite of it being bleak, this teaching can sometimes be experienced as good news. "You mean I am not the only person who is dissatisfied with life? I thought everyone else had it together and that there was something wrong with me. Now you tell me that we all feel that way. What a relief." Other words continue to appear; anxiety and trauma are sometimes used in their place.

Dissatisfaction and discomfort, according to the Buddha, is the nature of the human experience. The fish impaled on the hook, the frog disappearing down the snake's throat, and the animal on its way to the slaughterhouse, all feel their own kind of pain and confusion. Human suffering is just one part of a vast cosmos of suffering.

2. Suffering comes from desire

The second observation of the Buddha is that this misery ultimately arises out of desire, sometimes translated as 'craving' or 'frustration'. Humans are dissatisfied with their lives because, for some reason, they always want something more. There is a fine meal and then disappointment because one is too full for dessert. We already have a lover and then find ourselves looking at another possible partner with desire. We buy a big house and find the roof leaks. We buy a new car and find it is already rusting. The process of frustration is inevitable because, fundamentally, all things are impermanent, and we don't want them to be. We want the perfect life, yet all the while the carpet is being pulled out from under our feet. Our baby grows up, our lover ages, the house falls apart, the car deteriorates, and our own bodies and minds weaken. At the end, we too will decay, and this fear of falling apart and disappearing can be seen as simply the flip side of desire. We want to live forever. This fear is so great that humans invent all sorts of busy activities and elaborate, even heroic stories just to keep our mind busy.

Much of the chronic busyness we see around us (or call it frenetic activity if you prefer) seems to be driven by attempts to hide from the First Noble Truth. A vicious cycle begins with craving some kind of permanence, and each cycle ends as something new falls apart and disappoints us. This cycle is known as samsara. The Buddha taught that all humans are entrapped in the vicious cycle of samsara. The more we struggle to make life permanent, safe, cozy, peaceful, and happy, the more we in fact generate confusion, pain, fear, struggle, and misery. If the Buddha had arrived in the twentieth century, he would have seen this impersonal process of samsara spinning along just as he left it in 500 BCE, except that we now have the internet, telephones, and airline travel to speed things up.

3. There is a way out

This proposition seems rather simple and perhaps unnecessary. But perhaps it was added because the above two propositions seem so negative that an audience might leave in disgust. Note, by the way, that it is entirely irrelevant whether we approve of the above two observations, just as it is irrelevant whether or not humans approve of the nuclear chain-reactions that fuel the sun or of the angular momentum that keeps Jupiter rotating. A rational process of enquiry, one with respect for the nature of the world as it is, does not ask whether or not our ape minds happen to like a theory or not; the only question to ask is whether it fits reality. Most people in their heart of hearts know each of the Buddha's axioms to be true, even if it is so unpleasant that we try to avoid thinking about it or hope for a deity of some sort to protect us. The Buddha insisted that we have the courage to examine the nature of dissatisfaction; this, he taught, offered a means of escape.

4. The way out is through meditation

Meditation offers a practice to allow us to repeatedly wake up to the present, rather than living within some sort of elaborate fairy tale about our own lives. Meditation, as the Buddha presented it, provides a technique where humans can sit still and watch the kaleidoscope of their mind in a detached manner. Even when sitting still on a comfortable cushion in an airy hall, surrounded by others, with meals prepared, the mind continues its game of generating illusory experiences that we both crave and fear. Sitting still on a cushion, relaxed, relating with our breath according to the instructions the Buddha provided, is an awakening experience because, for the very first time, it is possible to experience that what is going on in the world of perception can be, and usually is, quite divorced from the illusions generated by our mind. One discovers that one wakes up again and again from the realm of illusion, only to fall back into the habitual thoughts and feelings. Therefore, the Buddha said, just take time to sit still and watch the process of mind in a gentle and detached way. Do not get caught up in the dream, for doing so only strengthens its hold on you. He taught students to gently practice detachment from their thoughts and to return to their body and their breathing whenever the illusion became too strong. Slowly this process empowers the tendency to wake up and weakens the momentum of the cycle of samsara. Slowly one begins to see the world as it is, rather than as a screen upon which we project our hopes and fears.

It soon becomes apparent that humans have a large streaming library in their minds, and something out of sight is busy playing those movies over and over again. One may be sitting on a cushion for only half an hour and one's mental processes are replaying any one of a dozen familiar but intense dramas. Perhaps one is having a love affair in Kathmandu, reliving a fight with a childhood bully, sitting down to a fine meal, worrying about a forthcoming speech, hating a former spouse, calculating one's bank balance, or just feeling irritated at having to sit still on a stupid cushion. The particular storyline of the movie is largely irrelevant; becoming familiar with the collection that is played is illuminating. Samsara is habitual. The emotions that drive the process may be general characteristics of apes, but each individual appears to have embroidered their own habitual stories around these emotions.

The Buddha put the practice of meditation at the heart of his instructions for students. Over his lifetime, he embedded them in a teaching structure called the Eightfold Path, which explains how to join our day-to-day living with the insights gained from meditation. One of the eight is, for example, the concept of *Right Livelihood*—finding employment that does not cause harm to others. Yes, the Buddha assumed that many students were actually going to have jobs, not beg for a living. In this book, we are mostly going to focus on Buddhist meditation itself, but if you consult other sources, you may see more references to the Eightfold Path. In Chapter 4, we will encounter a contemporary Buddhist teacher called Bhikkhu Bodhi and his book on meditation called *The Noble Eightfold Path*. In Chapter 10, I will briefly outline the Eightfold Path itself but, for most of the rest of this book, we will focus on the heart of the matter—the practice of meditation.

1
Craving for Resources:
Desire, Dissatisfaction, and Suffering

> I can't get no satisfaction.
> ("Satisfaction," *The Rolling Stones*)

JUST WHY WAS IT that the Rolling Stones couldn't get no satisfaction? After all, they lived in one of the wealthiest consumer societies the world had seen, a society so technologically advanced that it became the first, and arguably the only, society to free its own slaves. Further, it was a society where consumption was not only a source of pleasure and an opportunity to display power and prestige but also a social duty to keep one's fellow citizens employed. The Rolling Stones themselves were wealthy beyond the dreams of many ordinary citizens. Yet there was no satisfaction.

The Buddha made the same observation nearly two-and-a-half millennia ago. The First Noble Truth, the foundation of all Buddhist teachings, is that life is 'unsatisfactory'. This is a translation of the Sanskrit word *dukkha*. The word 'unsatisfactory' is usually expressed with a slightly bleaker English word and rendered as 'life is suffering'. However, the Buddha was not just talking about toothaches, heartaches, and disease, he was addressing something existential: the annoying, irritating, ever-present sense that whatever we are experiencing, it is not enough. Stephen Batchelor, a former Buddhist monk, suggests an even more intense translation for dukkha: anguish.

Packing for a wilderness hiking trip vividly teaches a similar lesson. Our wants vastly exceed our needs. Everything you will use on a trip lasting a week (or more) must fit in one small pack. And, you will daily feel the weight of each and every item you carry. We surround ourselves by so much clutter in our daily lives, trying to find some satisfaction. Hence, even before you arrive for the first day on the trail with Darwin and the

Dukkha is the Sanskrit word for a fundamental human experience. It is usually translated as 'suffering'. Hence, "life is suffering." More nuanced translations include: dissatisfaction, anxiety, irritation, or even anguish. This unhappy state of mind arises out of craving.

Buddha, you will have to look carefully at your own mind. Just what do you *need*? If you allow yourself to be greedy, you will be reminded by the weight on your back, every hour of every day along the trail. A tin of canned beef might make a luxury treat part way through a long hike, but you will carry that tin the entire way. We leave no trash. Crowding on wilderness trails means that the days when you could simply bury your waste along the trail are mostly gone. (Most of our cities still cannot cope with the waste they produce. Even the holy Ganges River has become one of the largest sources of discarded plastic debris entering the Pacific Ocean.) Craving, consumption, waste, and the sheer pain of material possessions all become vividly clear while preparing for a hiking expedition.

The Buddha's explanation for our chronic sense of misery is that craving has become habitual. This habitual craving prevents us from experiencing satisfaction. As time flows, moment-to-moment, we are tormented by visions of what we want, as opposed to simply appreciating what actually is. Sometimes we think we can escape by over-doing self-denial: ascetic practices, such as starvation, exposure, and more extreme forms of self-abuse sometimes used in the Buddha's time, and still used, in an attempt to overpower this craving. Ascetics might, for example, display their victory over craving by holding one arm erect until it became rigidly fixed there and withered. The logic was that if the mind could gain power over the body's insistent cravings, liberation would follow. Of course, such extremes of self-denial and self-abuse might threaten the existence of the body ... but this was believed to be the price one had to pay for true liberation.

The Buddha encountered these extreme ascetic views in his early training and explored them for some years. We may be certain that the desire to escape craving predates him because he inherited ascetic and meditative disciplines that were already available in India. Tradition tells us that the Buddha became enlightened only when he relaxed these strict practices, took some food, and sat under a tree in the shade. These stories about the Buddha are relevant to our inquiry because they re-emphasize the Buddhist view that craving is no small matter. It is *the* primary cause of misery and confusion in human beings.

But why do we crave? The traditional Buddhist view is that we should not get sidetracked into this kind of intellectual speculation. If there is a poisoned arrow in your heart, we are told, pull it out and treat the wound! To speculate on who made the arrow and what kind of wood was used will only distract us from dealing with the immediate problem at hand. Therefore, return your attention gently back to the experience of the arrow itself.

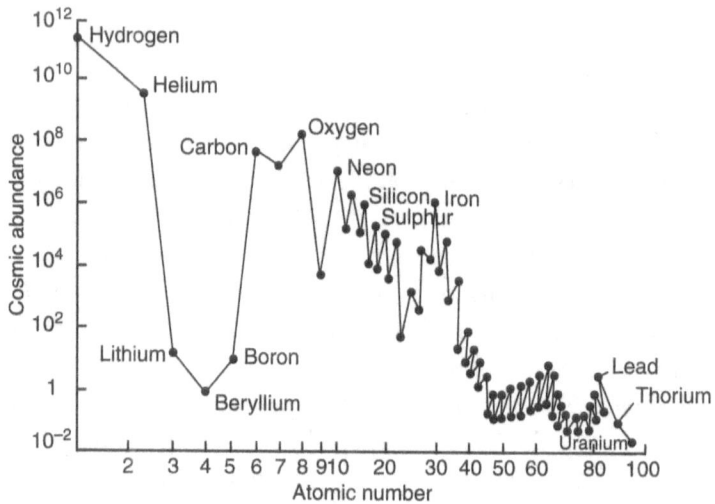

The chemical composition of the universe is very different from the chemical composition of living beings. All living beings must therefore continuously struggle to acquire and concentrate needed atoms and molecules from their surroundings.

Darwin (and indeed the western intellectual tradition), however, challenges us to explain how human minds have become inflamed by this particular source of misery. Why does this particular psychological state, rather than some other, torment us? And why is it apparently insatiable? Any enquiry into the origin of psychological states must, like inquiries regarding the opposable thumb, binocular vision, or erect posture, take us back in evolutionary time, when humans lived in small tribes foraging in the savannas of Africa. Or, even further back. Some of the physiological systems in human brains can be traced back to oceanic ancestors.

Suffering, evolution and membranes – a biological foundation of craving

Evolution by means of natural selection does not have a goal and certainly does not necessarily act to make people happy or content, unless happiness and contentment somehow enhance survival and reproduction. The numbers of genes in the next generation is the primary means by which evolution acts. If misery and craving enhance survival and reproduction, there is every reason to expect them to increase in future generations. We might therefore begin with the assumption that human brains produce enough craving to ensure survival and reproductive success, whatever the consequences for our happiness. Craving would decline through time if its costs exceeded its benefits, as measurable in terms of survival or reproduction.

If, say, craving became so extreme that we killed ourselves or destroyed our families, then over time the propensity to crave might decline in intensity. So let us look at the biological origins of craving and consider the costs and benefits.

Craving may be one of the most ancient attributes of animal life. Its origin almost certainly lies not with early human cultures, nor with tribal apes, nor with arboreal primates, nor with mammals, not even with vertebrates. *Craving, rather, is a psychological manifestation of a chemical reality of life on Earth*: all living organisms are built out of a small subset of the elements and molecules found in the universe, and these elements are scarce. Hence, all living organisms live in a constant struggle to extract needed elements from an environment that is deficient in them. As soon as collection of these resources ceases, organisms begin a descending spiral toward death.

Every organism in every environment, then, is constantly struggling to accumulate the necessary resources for its survival. Freshwater fish are struggling to extract calcium, sodium, and oxygen from the water around them. Saltwater fish are struggling to extract water and oxygen from the saline solution around them. Plants are struggling to extract water, nitrogen, and phosphorous from the soil around them. Even the humble bracket fungus on a log may have many kilometers of thread-like hyphae struggling to dissolve resources out of the tissues that were once a tree. The widespread human preference for meat also has a biological basis: the best source of key nutrients, including nitrogen and phosphorus, is the body tissues of other living organisms.

Throughout nature—in all living beings—there are systems of semi-permeable membranes and ion pumps running day and night. These distinguish between the interior and exterior of cells, and they move elements from the outside to the inside. (Others discharge waste products, but that is another story. Here we are concentrating only on the uptake of raw materials.)

Even in your own body, the bacteria (and possibly the tapeworms or roundworms) in your intestines are struggling right now to remove essential elements from your food before the food is absorbed by neighbouring bacteria or the cells lining your intestines. The effort with which an organism is struggling to take up nutrients can be measured by its rate of respiration; the harder it is working, the more oxygen it consumes, and the more carbon dioxide it releases.

It can therefore be said with certainty that this kind of struggle originated with the earliest biological membranes. A membrane is essential for there to be an inside and an outside, an organism and an environment, an individual and a community. As soon as a membrane was present, there was the possibility, indeed the necessity, for molecules to accumulate on one side or the other. As soon as the earliest membrane-bound container evolved, perhaps some two to three billion years ago, the struggle to accumulate resources began, and it has continued unceasingly through time to the present. What we call the experience of craving is at least, in part, a conscious expression of the imperative to accumulate resources. In this sense, craving is just another tool, like beaks, eyes, claws, or gills, which assists us in acquiring resources.

The unconscious foundation

The breadth and depth of life's craving for resources must be fully appreciated. Resource acquisition is one of the dominant themes in the discipline of ecology. Consider nitrogen, a resource required for constructing proteins that form the bodies of plants, animals, and fungi. Indeed, it might not be stretching the truth too far to state that the Rolling Stones were describing the state of perennial nitrogen deficiency encountered by life forms on Earth.

"Water, water everywhere, but not a drop to drink." We can pity the poor ship-wrecked sailor, who, while dying of thirst at sea, cannot drink the abundant seawater. Similarly, humans use nitrogen to build protein, and the atmosphere around us is roughly three-quarters nitrogen. However, we cannot use that nitrogen and must instead obtain it from other living beings. We can get nitrogen from plants, but all plants have the same problem we do. Most plants, like us, cannot extract nitrogen from the air. The corn plants of the American mid-west that feed much of our world are immersed in an inexhaustible supply of atmospheric nitrogen, yet farmers must provide nitrogen fertilizer. This application of fertilizer is, in fact, one of the most expensive parts of corn production. To understand why this problem arises, one must go back some three billion years to a time when there was no oxygen on Earth. But first, a little more on nitrogen.

Nitrogen is the gold standard, the US dollar, and the Swiss bank account of the ecological economy. It is valued by life above all other elements and carefully stored for future use. Nitrogen is essential for building molecules to construct the basic structures found in plant and animal bodies; proteins, neurotransmitters, chlorophyll, DNA, and RNA all require nitrogen. Since animals depend upon plants to provide them with nitrogen, animals are limited by nitrogen supplies just as much as plants. Here is a practical example. The list of ingredients on any bag of fertilizer reveals that nitrogen is a major component. Just like plants, a gardener is paying for an element that bathes every leaf and stem in the garden. How can this be?

Nitrogen in the air is in the form of an inert gas. Plants, for reasons that are still unclear (partly because they are shrouded in the mists of the early origins of life), cannot use this nitrogen. They instead take up nitrogen in other chemical forms, particularly ammonia, as well as nitrites and nitrates. The dependence upon ammonia ions may be particularly significant since, so far as we can tell, ammonia was common in the Earth's early atmosphere. At that time, a cellular metabolism depending upon ammonia was presumably a quite reasonable evolutionary proposition. Further evidence for the evolutionary primacy of ammonia is the discovery that when plants take up nitrogen in the form of nitrate, the nitrate ions are immediately reduced back into ammonium within the cell.

One way to measure just how desperate plants are for nitrogen is to look at the bizarre means that have evolved to circumvent this shortage; in a similar way, the desperation of addicts can be measured by the extremity of the means they are willing to employ for the next fix. Carnivorous plants capture the imagination in the same way that newspaper editors like the headline "Man bites dog." There is something faintly satisfying about the prospect of plants getting even after millennia of being eaten by animals. Pitcher plants, sundews, Venus flytraps, butterworts, and bladderworts, although they occur in different plant families, all share two traits: (1) these plants occupy particularly infertile habitats such as peat bogs, (2) animal proteins are used to supplement the nitrogen economy of the plants. Carnivorous plants, then, are an extreme form of adaptation to infertile environments. They illustrate the intensity of the struggle to capture resources from an unforgiving world. (Darwin experimented with the diet of these plants in his own greenhouse and eventually wrote an entire book on the topic, published in 1875, *Insectivorous Plants*.)

Nitrogen is the key element in the story. Nearly all higher life is dependent upon a small group of living organisms that can build protein using atmospheric nitrogen. An ancient group of micro-organisms, cyanobacteria (once called blue-green algae), have retained the ability to remove nitrogen from the atmosphere. These cyanobacteria are true living fossils. Their ancestry apparently goes back some three billion years, before life left the oceans, before there was oxygen in the atmosphere, and before there were fish swimming in the sea. Today they live in marshes and swamps or sometimes within the tissue of other plants. They can thrive only where oxygen levels are low which, of course, reflects the situation early in the Earth's history.

It is subtle and outrageous, yet true: without these ancient cyanophytes, these living fossils from a time when Earth had almost no oxygen and no higher life forms, the fertility of soils would steadily decline. The only other natural process that returns nitrogen to soils is lightning discharges. All higher plants have been able to depend upon cyanobacteria and lightning to replenish the soil with nitrogen. All animals contain nitrogen from plant tissues that was once fixed by cyanobacteria and lightning. We may all, then, be bathed continually in atmospheric nitrogen, but it must squeeze through a tiny bottleneck, the cyanobacteria and lightning. Similar stories can be told for other essential elements such as phosphorus and sulphur.

In summary, the existence of life is predicated upon resources to build and reproduce bodies. The extent to which all life is in a desperate struggle for nitrogen illustrates the way in which life is constantly limited by shortages of essential resources. Membranes provide the tool to accumulate these resources. It is therefore hardly surprising that we should experience distress when the supply of resources is limited.

Costs and benefits

It is no wonder that we crave. Our minds are responsible for maintaining bodies that must accumulate resources. Therefore, however much discomfort is caused by craving, it has always been essential for animal survival. Craving impelled our ancestors to find new sources of food, to hunt down one extra rabbit for the pot, as well as to seduce one more mate and to generally enhance their genetic contribution to the next generation. These same ancient genes still govern our physiology and behaviour. The more we accumulate, the faster we can grow, and the more prolific our reproduction.

The existence of craving is therefore not such a puzzle. More puzzling, perhaps, and certainly more relevant to the issue of human suffering, is the observation that there seems to be no natural limit on, or end to, this craving. Give us a fine house, a freezer full of food, a large family ... and we are still dissatisfied. We continue to crave even when logic dictates that there is nothing more to crave. The displays of extravagant consumption by kings, emperors, popes, tyrants, and dictators illustrate what can happen when craving is entirely unleashed.

Why does there appear to be no upper limit? For there to be such a limit, there would have to be costs to craving. If craving had benefits and no costs, it would steadily increase through time. Is there any natural process that might set an upper limit to the amount of craving we inherit? Misery and dissatisfaction are not considered to be real costs unless they influence survival or reproduction. As long as we are not so miserable that we kill ourselves or become celibate, the costs are few. Throughout human history, however, we read of suicide and of people who were sufficiently unable to cope with their social environment that they failed to reproduce. Perhaps suicide and reproductive failure are both outcomes of the craving thermostat begin turned up too high.

If we turn to Shakespeare, we might take note of the number of his plays that turn on envy and ambition. Both of these emotions can be regarded as specialized states of craving. Envy is the desire to acquire something that belongs to someone else (sometimes the result is pillage and plunder, Chapter 6). Ambition is the desire for another limited resource, stature in the community, which is usually accompanied by wealth and prestige (often the result is hierarchies, Chapter 5). Let us say something more about status: in many groups of animals there is a strong correlation between the reproductive success of a male and his status within the group, so the struggle for status may be regarded as a type of craving that goes back well into our past when our ancestors first formed social units. Further, since social status still confers better access to resources, ranging from food to medical care, and more resources to allocate to children (including illegitimate children of mistresses), social status is still likely correlated with reproductive success in our current era. But now we are looking for costs that would counteract these benefits. Does Shakespeare give us any clues? We may read his plays for a simple empirical description of the consequences of human vices.

Viewing *Julius Caesar* from this perspective, the craving of Cassius and Brutus leads to their premature deaths. Brutus says:

> Into what dangers would you lead me, Cassius,
> That you would have me seek into myself
> That which is not in me? (Act I, Scene I, l. 63-65)

Humans may also have learned to instinctively mistrust people who were too evidently filled with craving. Caesar says:

> Yond Cassius has a lean and hungry look;
> He thinks too much: such men are dangerous.
> (Act I. Scene I, l. 194-195)

Similarly, the chaos in the play *King Lear* is created by an unwise allocation of resources by a father, while the tragedy of *Macbeth* is driven by the ambition of Lady Macbeth. In short, within the plays of Shakespeare, we may see evolutionary shadows of the costs of craving. These costs are social disorder, which disrupts one's rise to power, reduces access to resources, and risks the survival of one's family. So a further cost of craving may have been the unease it arouses in others which might lead to a reduction in one's chances of survival and reproductive output.

Apart from these, there may be few other costs to craving. The number of people wealthy enough to eat themselves to death has been few and far between (the contemporary United States being a notable, but probably quite temporary, exception). Further, most of our most obese fellow citizens have already reproduced, so their death from heart attacks at fifty does not affect their reproduction; indeed, it speeds up the transmission of their financial resources to their offspring. We could even advance the argument that the inherited craving for the early and rapid acquisition of resources might advance the genetic prospects of their offspring. It is therefore possible that the amount of craving we each experience is set by the balance between its benefits (the necessity of acquiring resources for reproduction) and its costs (the misery that might lead to suicide and the risks from conflict with other members of our own species). These are issues to consider on your next grocery excursion, which will provide abundant opportunity to observe large numbers of people responding to their craving for resources.

Whether or not there are costs to craving, its existence in our lives cannot be ignored. We cannot live more than a few days without essential resources and, usually within a few hours of consumption, we already find the pangs of hunger calling our attention to the search for food again. This connection between hunger and craving has long been understood; note that Cassius had a lean and hungry look, but it is doubtful that Caesar or Shakespeare was talking about him needing a good supper.

Perhaps, then, we might feel compassion for other human beings who are trapped by craving. Traditional Buddhist teachings describe a spirit realm populated by creatures called hungry ghosts. These are imaginary creatures with huge, swollen bellies and scrawny necks, beings that are tormented by hunger they cannot satisfy. One contemporary view is that such beings represent a psychological state that afflicts humans. In this sense, you can see hungry ghosts all around you: in shopping malls, on tropical beaches, on surfboards, at bingo games, and, yes, even at meditation retreats and scientific meetings. We can say it was one of the Buddha's great insights into the human condition to have invented such a powerful image of insatiable desires. It is an experience that transcends humans. The porcupine chewing the bark off a red oak tree, the chimpanzee using a stick to spear a bush baby in a hollow tree, and countless other kinds of wild creatures may feel much like we do.

Craving has global consequences

At the global scale, it is important to understand just how human craving is fuelling the destruction of wild nature. We are now in the final

$$\left(\begin{array}{c}\text{resource use}\\\text{per person}\end{array}\right) \times \left(\begin{array}{c}\text{number of}\\\text{people}\end{array}\right) = \begin{array}{c}\text{total impact}\\\text{upon}\\\text{Earth}\end{array}$$

The impact of humans upon the Earth in general is driven by a simple equation. There are two terms. The first is the impact of each individual, which is determined by the resources consumed by that individual over their lifetime. The second is the total number of individuals. Both terms in this equation are driving wild nature to destruction. Too often conversations about 'the environment' focus on only one-half of the equation. It is entirely possible to have both harmful terms acting simultaneously.

stages of turning wild nature into human bodies. It is that simple. In this search for food, forests are turned into monoculture soybean fields, wild animals are shot into extinction for bushmeat, and the oceans are emptied of fish by giant floating factories. This process is, of course, unsustainable, which is why we have a large and growing list of endangered species. Scientists maintain an International Red Book of endangered species. There are more than 26,500 species in that book; you can read about each one at https://www.iucnredlist.org.

How is this happening? There are just two causes. The natural world is being destroyed by a machine with two huge engines running in parallel. Each engine is fuelled by craving. The first engine is fuelled by craving for resources or, more generally, material items. This is particularly obvious in the developed world, where even people who seem to have everything they could possibly want still crave an expensive new car or a cabin cruiser or a bigger house or a new granite countertop. Cars, yachts, and houses are made out of natural resources somewhere, and hence, the more you consume, the more destruction you cause. Even existing parks in Southeast Asia, set up to protect tropical rainforest, are being ravaged by illegal logging driven by the market for wood used to make expensive consumer goods. Of course, it is easy to condemn the rich and their lifestyles, but it is part of the story.

The other engine destroying the Earth is craving for reproduction, the drive to leave more copies of ourselves. Darwin had a lot to say about that. Even in areas of the world that cannot support their existing populations, people continue to have many more children than their land can support. We repeatedly hear about famine in Africa, yet after each crisis passes, the population increases again. Consider just one country known for repeated famines in my lifetime. Since 1960, the population of Somalia has tripled! It has a climate and landscape that is inherently prone to desertification. Every baby is a new replicating unit that must turn wild nature into its own flesh. Since there are regions of Africa that can no longer support their current human population, boats full of migrants are exported to Europe, while planeloads of migrants land in North America. Of course, if we try to give every baby who is born an SUV, a large house, and a cabin cruiser, the result will be disaster. You can watch the results of this craving to make babies in real time at http://www.worldometers.info/world-population. In the time it has taken to read this chapter, hundreds of new human beings have been born into poverty, and each will crave its own share of what is left.

Mostly I find that the poor blame the rich for their conspicuous consumption, while the rich blame the poor for having too many babies. This seems to be a good way of avoiding the reality that both are more than the planet can bear and that both are an unhappy consequence of craving.

When we are on the trail, we come face to face with our craving. We pass through feelings of fatigue, hunger, thirst, irritation, and many more mental states. We also experience magnificent views of forest and mountains, and we meet countless other living beings. Some of those beings sing outside our tent in the morning, others rustle in the leaves around our tent at night. We notice our bodies too: it feels really good to put down our pack and take a rest. It can also be painful to lift it onto our back and start walking again, but it is choiceless. The end of the trail is the only way out. Discomfort is simply part of the journey. The hours on the trail offer time to reflect. Did we really need that luxury swiss army knife with the extra blades? A smaller one would have weighed less, and likely still have met our needs.

Coming back to the First Noble Truth, the uncomfortable feeling we call craving arises in each of us, continually, whether we are on the trail, at the office, or in our bed. It is a part of our natural heritage. We are here because our ancestors craved natural resources and craved babies. We don't condemn ourselves for having binocular vision, nor for having opposable thumbs. They are something we learn to live with. Craving is part of this biological heritage. At the same time, just because the urge to eat and the urge to reproduce are 'natural', it does not mean that either of them is, therefore, inherently 'good' or even 'fair'. Any hope for a sustainable world requires us to engage the world in a realistic way, to engage with what is, rather than what we want. It appears that humans really can't get no satisfaction. Like Cassius, we all have a lean and hungry look about us.

2
Living an Illusion: Mind as Cocoon

> You can check out any time you like,
> but you can never leave.
> ("Hotel California," *The Eagles*)

WE RARELY SEE the world around us as it really is. Yet we assume that we do. Most of the time, we are viewing our world though a cloud of our own emotions, and, to make it worse, we do not even realize that this is happening. That is an unhappy mixture—confusion made worse by ignorance. This is one way that craving subtly controls our behaviour. Even our most basic five senses can be clouded by hopes and fears, or simply ignored. Hence, we may meet new people, and we may experience new locations, but we are always seeing them partially obscured by various emotional states. We are conditioned by previous experience. Our mind is often so cluttered that we make mistakes in interpreting what is happening in our lives. A simple but familiar example is eating a meal, only to discover at the end of it that we have not tasted the food because our mind was chattering about office politics. Or, we meet a co-worker and feel so angry at their past behaviour that we cannot really hear what they are telling us about today's project. Or, we end up in an argument with a spouse, only to discover that we do not remember what set it off. How can we possibly expect to have a decent life if we cannot clearly perceive our surroundings? How can we communicate with other people if they too are trapped in similar misperceptions? How can we check out of Hotel California at all?

The Buddha taught about this problem several millennia ago. It is still relevant today because he was talking about how people experience their own minds. Now we have study after study of human perception showing that much of what we experience during a simple lunch with a friend or a short walk outside is not so much a film but rather a hologram. Perhaps even a hologram that is carelessly constructed. Our view of the world is constructed from selected perceptions, combined with selected memories, and then distorted with certain expectations. Modern advertising and propaganda are using such information to further manipulate our view of reality. Thus it is helpful for us to study our own minds, with an emphasis upon how such confusion arises.

We could say that the entire path laid out by the Buddha had the simple goal of clarifying our relationship with reality. There are many books and teachers to assist us with this endeavour. But here I want us to look more carefully at the source of our confusion itself. How is it that our ape mind is so determined to create its own version of reality? The point seems to be that every emotional state has a biological purpose. Indeed, even the tools of perception themselves, like our eyes and our ears, have a long evolutionary history. Darwin wrote at length about how animals in general, and humans in particular, came to be the way we are. There is nothing like carrying a pack along a mountain trail to make us painfully aware that we have an intertwined physical and a mental existence. The mountains are there. Our bodies feel exertion and pain. Our senses show us birds, plants, and rocks. And all the while our mind wanders, sometimes noticing our sense perceptions, but frequently telling us elaborate stories. It becomes almost a game: when are we really on the trail, being with our companions on the hike, and when are we wandering through the inner world of thoughts and illusions? Sometimes you get stuck with a companion who can't stop complaining about their life back in the city; they do not seem to be on the trail at all.

So let us leap to the big question: why should consciousness exist at all? Why do we perceive and feel our existence in the way we do rather than some other way? Most of the world's religious, spiritual, and psychological books are concerned with such questions. Many, but not all, short-circuit the discussion with the answer that a creator chose to make it the way it is … period.

Let us begin instead with a Darwinian view. We have already seen that craving for resources is a fundamental process in the natural world and therefore a perfectly natural part of our own psyche. So the First Noble Truth of human suffering is really just an empirical observation of how all organisms relate to their surroundings. Organisms need not be conscious of their craving and therefore experience it as suffering. It seems that some groups of organisms, like marine invertebrates, probably function rather well with rather low levels of consciousness. Yet, even humble barnacles use serotonin and dopamine to make decisions about where to settle and grow. These are the same molecules that our own brain employs to make assessments of our circumstances. One important feature of humans, however, is our capacity to consciously experience those circumstances. Of course, we don't always feel the craving consciously; much of the time it is relegated to the unconscious, or to the edge of subconscious. The human mind has the capacity to move our feelings in and out of our conscious awareness. At some level, we are filled with a variety of emotions, and our awareness deals with them one at a time as they arise. Why should it be this way?

Darwin's insight into our predicament was this: we are here only because our ancestors persisted long enough to reproduce themselves and, more precisely, because they did so more successfully than others around them. What we carry forward as consciousness exists because our ancestors bequeathed it to us. It worked for them, more or less, and now we have inherited this style of consciousness. Those who died young, or did not have offspring, did not leave their experience of the world to us. This is self-evident. It is true by definition. Yet it is profound in its implications. Each of us is here because of an unbroken chain of survival and reproduction going back far beyond our own lifetime. To overstate the case somewhat, reptiles have scales, birds have feathers, mammals have fur, and *Homo sapiens* have consciousness. While we may not understand the details, we know that we are the end point in a process and have inherited our experiences from ancestors who inherited theirs in a similar way.

This, of course, is a rather superficial introduction to the topic of consciousness, but I am not trying to summarize all that great psychologists, religious leaders, and philosophers have said about consciousness, and free will, and all of that. You can read that elsewhere. Here I want us to leap from Darwin to the Buddha. Darwin showed us how human consciousness originated in our ancestors. Where do we turn for more descriptive data of what human consciousness is actually like? Here we have a treasure trove: for 2,500 years Buddhist meditators have systemically and meticulously described every aspect of human consciousness. Thoughts, feelings, impulses—strong ones and weaker ones—are extensively documented in the Buddhist literature. So we have a large body of empirical data on what people actually experience. It is a kind of data archive on the natural history of the human psyche. Let us look more at this rich pool of data.

The process of data collection and classification of the inner world by Buddhists has been every bit as systematic and scientific as the classification of birds by biologists. The existence of 9,750 species of birds in the world, for example, is the result of hundreds of years of patient observation by scientists and amateurs around the globe. No one has seen all of them. But by collecting observations and combining them and relating them one to another, we can come up with an encyclopedic description of the world's birds. Each special place has such a list: the Great Smoky Mountains has approximately 120 species of breeding birds. Some, like the ruby-throated hummingbird, also nest on my own property in Canada, but others, like the hooded warbler do not breed so far north. I hope to see a hooded warbler on this hike. We will not see a Bicknell's thrush, as they breed much further north still, but they will show up in the conversation, later. In the same way, the observations collected by skilled meditators over thousands of years provide similar compendia of the human states of consciousness. These observations are also unified with some empirical theories. Just as we can turn to a scientific library for a description of the birds of the world, we can turn to Sanskrit or Pali documents for descriptions of the elements of human consciousness. The Pali documents, for example, describe no less than 52 mental factors than can be present in consciousness. Most of us live our lives without noticing most of them, in part because we are not paying attention, in part because we lack the appropriate vocabulary, and in part because we lack experience in mediation. In the same way, when the average human being walks through a forest, they see few of the kinds of birds there too. In the same way, when the average human being experiences consciousness, most of the details remain unseen. Just as the warblers and vireos are there but unseen in the forest, *phasso* and *panna* and the other fifty factors of consciousness are there but unseen in the mind. They can be sorted into five *skandhas* (factors involved with craving), twelve *nidanas* (a series of cause and effect linkages), and other systems, but they vividly describe the moment-to-moment experience of consciousness.

What about personal experience? It is only (or mostly) when we sit still in meditation that we can begin to experience the movie of consciousness as being composed of individual frames, each frame in turn containing scenes and meanings, each scene being assembled from perception and feeling. Some humans will sit in meditation and experience phasso and panna and the other elements of consciousness for themselves. Similarly, some humans will make the trek to Trinidad to see trogons and toucans. Many others will not see them except in books. For too many humans, the experience of life as it arises is taken for granted, and, what is worse, experienced in a confused way. This confusion causes us to suffer. We are actually living our lives out of touch with reality, enclosed in a cocoon of our own thoughts and emotions.

From the Darwinian point of view, these elements of human consciousness exist because they enhanced the survival of our ancestors. From this perspective, consciousness is a simulation model of reality that is constructed largely from sensation and memory. The sense organs gather information about factors such as the prevailing light conditions, chemical concentrations, temperature, and other surrounding organisms. The brain then uses these perceptions, along with stored memories of similar conditions, to create a model of the world. We then place ourselves within the model. Those familiar with Buddhism will know that some of the Buddha's most profound insights related to the experience of the self and the realization that the self in the model is an *illusion*. It is not real. According to the Four Noble Truths, our suffering arises because we believe this simulated self to be real. It is not, and the *Heart Sutra*, which is recited each morning in some Buddhist traditions, recounts how shocked people were when they first heard the statement that our experience is illusory (or empty). The *Heart Sutra* also identifies the location of this event: Vulture Peak Mountain, which you can still visit in modern India. Neither Darwin nor the Buddha played video games, but if we are looking for a modern example, we might say that to believe that our self is real is the same kind of mistake as believing that our character in a video game is real.

But we are getting ahead of ourselves. (Note how the English language assumes that each of us exists, and doubles it for emphasis—with two words *our* and *self*.) We will look more at this illusory self in the next chapter. The point here is that our interior model of 'world' and 'self', and the forecasts derived from it, allow an organism to respond to the world with appropriate behaviour. For example, if the brain reports that a potential mate is near and if the brain reports that certain criteria are met, the feeling of attraction is generated, and this leads an organism to approach a potential mate. If this sounds overly analytical, you were not paying attention the last time you attended a social gathering of humans. Men are generally attracted by women. The degree to which this simple model drives human behaviour is the theme of philosophers, humorists, and every soap opera on television. You see it in shopping malls and at elite scientific conferences.

Equally, if the brain instead reports a dangerous predator, the organism is impelled through fear to move away. Attraction and aversion: without any more sophistication, we have now encountered two of the most powerful forces underlying human consciousness. If you doubt the biological basis of these two emotions and the power of natural selection in driving them, just imagine your chances of survival if the model used by your brain contained a single error. Imagine if it instead generated sexual desire at the sight of a predator and fear at the sight of a potential mate. How would that work out? A person with that misguided model of reality would not likely survive long enough or mate often enough to leave any offspring.

The title of this chapter uses the word cocoon, which is a term introduced by Chögyam Trungpa to describe how we live within our own mental illusion. Some aspects of this illusion may be relatively accurate, some less so, and some probably completely wrong. But we carry this package around with us and assume it represents reality. In fact, this cocoon shields us from reality. So, if we want to understand our reality, we need to have the courage to look at this cocoon, and then we need to rouse the courage to look at the world as it is outside this cocoon. This process is not just the Buddhist path; Chögyam Trungpa generalized it further by identifying is as the Sacred Path of Warriorship. He used the word warrior deliberately because the process of dealing with our cocoon demands a certain amount of courage. He also emphasized that this path is a secular path. We do not need to put on monastic robes or even call ourselves Buddhists.

The cocoon, then, is deeply connected with our felt experience of life. In *Touching Enlightenment*, Reggie Ray describes how the cocoon has some short-term benefits. It acts as a needed filter upon our relationship with the world. He notes that the world provides enormous amounts of sensory input, so much input that we simply cannot process it all with the intellect. Hence, part of the function of mind is to filter out sensory input, delivering only selected information to our conscious awareness.

Here is an everyday example of this process. When I take visitors into our forest, I am frequently struck by the number of them who do not see birds or even porcupines sitting in the trees, unless I take the time to point them out. And for anyone who has never tried to point out a wild bird sitting in plain view, you would be astonished at just how difficult it can be for people to see what is right in front of their face. Our ability to see our surroundings is so heavily filtered. Reggie Ray describes our spiritual path as being both able and willing to become aware of the material that the filters have removed. He also believes that much of this material is actually captured and stored, but simply not delivered to our awareness. Long periods of meditation may allow us to recover these unprocessed events.

Reggie Ray, of course, was a professor of religious studies, not a professor of biology. However, he does provide a vivid account of how the cocoon affects our perceptions and how we might work with it as an experience.

These views of consciousness are consistent with what we know of evolutionary biology. We have inherited a system that processes our sensory input, selects certain aspects of that input for awareness, and stores other aspects away for future use. It also may generate an appropriate response to the input by creating an emotion. Here I use the word 'appropriate' in an evolutionary context. Too many people these days do not understand that there is a difference between an appropriate evolutionary response and an appropriate social response. When we cross the room to introduce ourselves to a potential mate, it is completely appropriate from the point of view of natural selection. However, it may be regarded as socially inappropriate if the potential mate does not want the attention. Just how men and women are supposed to meet without causing offense is a topic of loud discussion these days, both in the popular press and even within my own Buddhist community. It is almost as if some of the commentators don't understand Darwin and human behaviour, or the role of eggs and sperm in nature.

To return to the theme of this chapter: the models projected within the cocoon help protect us from hostile conditions by guiding our behaviour to increase the odds of survival and enhance the odds for reproduction. And we can thank, or blame, our ancestors. Those ancestors whose models were accurate enough to drive the right decisions about predators and mates have bequeathed those models to us and we now use them.

More on illusion

The cocoon filters much of our input, determines what we will 'experience', and thereby elicits basic responses. Hence, our experience of the world around us will always be somewhat incomplete and considerably biased, like our visitors who walk through the forest and cannot see the birds. Now here is another consideration. Nowhere is it written that these inherited models have to give us accurate descriptions of our outer world. Models exist to elicit certain kinds of behaviour. Somehow, we have come to naïvely believe that the model of reality we experience is reality itself. It is easy to assume that, as our brains became larger and we became smarter, the accuracy of our mental models necessarily increased. We begin to imagine that there was progress. Progress is a dangerous word since it often leads to self-delusion. As Robert Trivers expressed it "...the conventional view that natural selection favours nervous systems that produce ever more accurate images of the world must be a very naïve view of mental evolution" (p. viii).

There may be strong advantages to deceit, including self-deceit. Bigger computers do not necessarily run more useful software; the added size may just enhance the vividness of illusory video games. (Indeed, it may be noteworthy that Darwin called his book *The Descent of Man* and that Buddhists talk about humans being surrounded by illusory projections that cause misery. Neither of these views assumes that progress is innately part of the human condition.) If false models and illusory descriptions of the world enhance survival or reproduction, then the minds that produce them will tend to multiply at the expense of others. Over time, the minds that predominate may lie to the organisms that possess them. If we look at consciousness from this perspective, we can see that there might well be survival advantages to brains that create convenient lies. Let us consider two of the more obvious lies we might be told.

More on illusions and warning

Some illusions produced by consciousness are almost certainly mere warnings, predictions of possibilities. Fear of darkness is likely a common example. Presumably it originated because predators killed any of our ancestors who lacked such a fear. Yet, on some nights, there might be no predators in the forest at all, and it would be quite safe for an ape to wander in the woods alone. The fear that keeps an ape in a cave, up a tree, or in a house, will protect it from those nights when real threats exist. Of course, it will also keep it inside on many nights when there are no predators, but that is the nature of forecasts: they deal with probabilities, not certainties.

There is a subtle balance at work in the construction of illusions. If the lie, if the illusion, if the story-line, were too obviously false, it might seem to be pointless or even self-defeating. If a drowning ape imagined land, or a thirsty ape saw visions of water, the visions would, at first inspection, not save them from the reality of death by drowning or death by dehydration. Imagine, however, that if a mental image of land were to encourage a drowning ape to swim harder and longer, this might, on average, lead to fewer apes drowning. Or perhaps mental images of water would impel thirsty apes to drag themselves over one more hill to an oasis. In such cases, the presence of tantalizing but false images would have real biological value. Minds that tantalize us with false images might therefore enhance survival and reproduction.

To return to the example of fear of darkness, we can also see how models need not be particularly accurate. In fact, there may be an advantage to having them vague, or even exaggerated. It is likely that predators were once a good reason to stay out of forests at night. But everyone reading this book knows that actually forests at night are full of monsters. The fear that really keeps us out of the forest at night—a fear we see in countless fairy tales, for example—does not accurately portray a saber-toothed cat, a cave bear, a jaguar, or a scorpion; rather, it deals with vaguely-defined impressions. A dimly imagined fantastic monster such as a werewolf, vampire, or ghost may provide a more powerful image to keep a wary ape inside than would a biologically realistic image of something specific such as a saber-toothed cat. There are real risks to being in the forest at night and, by the way, not just predators. There are also scorpions, disease-carrying insects, and simply the risk of falling over a log or having a sharp branch poke out an eye. Hence, fear of the darkness and fear of forests may summarize the risks from many kinds of danger, and the most powerful (and general) way of doing so may be with grossly-exaggerated images of mythological beasts, ghosts, and other assorted undesirables. We can experience this personally as well as intellectually. Even though I know my own forest intimately, and even I know that the dangerous predators are locally extinct, as dusk approaches I can feel a visceral fear rising, warning me that it is time to get home and sit beside the wood stove. My intellect feels slightly embarrassed by the experience, but then I remind myself that I am merely having a direct encounter with a neurological pathway that I have inherited from my primate ancestors, at least those who survived by staying out of dark forests at night.

Even real lies can have an evolutionary explanation

We may at first find it difficult to accept that minds do not strive for accuracy so much as effectiveness, but certainly the example of fear of darkness is one I find compelling. When I am alone in the woods, it is a far deeper fear than an image of a wolf or bear arouses that haunts me. It is the undefined nature of the fear, like the monster in *Beowulf*, which seems to make it worse.

We can likely imagine many other kinds of lies our minds might tell us. Speaking on behalf of the male gender, I can assure you that there really is a part of our mind advising us that we should go and introduce ourselves to strange women, and that if we do so, they will find us fascinating (indeed, says the unconscious, they will take you home to pass on your genes). The American comedian Bill Burr says that whenever he thoughtfully asks himself whether he should approach a female and listens respectfully for a reply, he hears, "Do it! Do it! Do it!" Now that is a real Darwinian joke. Our minds may lie to us not only about mythological monsters in forests, but equally about our attractiveness to potential mates.

But this line of thinking need not stop here. How elaborate is the illusion? Mental lies—projections, visions, hopes, fears—might be advantageous *even* if they had no immediate contribution to survival from obvious environmental threats. That is, even if the logic above is wrong, even if images of monsters did not keep children safely in their homes, even if images of death did not keep us away from cliffs, our minds might still mislead us about the external world around us. This is because all forecasts have an element of uncertainty. We can think of our fears as also being slightly exaggerated forecasts or predictions. They present us with the equivalent of a war game, or a computer-generated economic scenario. They warn or inspire us by presenting a possible result (being killed by a night monster) of a possible course of action (going out at night). To this extent, they are not so much illusions as they are warnings. They are the biological equivalent of a bright orange sign that glows in the dark and warns 'danger ahead'. The sign itself is not the dead end or the narrow bridge; it is a merely a useful and attention-grabbing symbol for a particular danger.

Illusions, projections, and imaginings could have one more function, however: that of preventing self-destruction. If an organism succumbs to pain, exhaustion, boredom, or despair and is driven by them to end its life (or to not reproduce), then the mind/body control system of the brain has failed. In this context, pain or mental despair leading to suicide is no different from a defective valve leading to heart failure. The purpose of consciousness is to sustain and propagate the body/mind container. If this requires lies, illusions, tranquilizers, or stimulants, so be it.

We therefore may not be able to trust our brains or our habitual minds because they are 'purposefully' set up by natural selection to delude and confuse us. The minds we have inherited are those whose illusions maximized rates of survival and reproduction in our unbroken lineage of ancestors; despair or pain might well have increased the former and reduced the latter. The brains that we have inherited might therefore be the least accurate. In this case, they exist not to help us discover truth in our lives (or to unravel the secrets of the universe) but rather to delude us into surviving and reproducing. The mind may not be a reliable ally in the search for truth or meaning. It may be just one of our parts, just like our stomach or our feet. Indeed, the very idea of 'truth' or 'meaning' might be one of the illusions, the ever-receding mirrors, the puffs of illusion, the threads that entangle us. From this perspective, we are surrounded by a special kind of membrane, a cocoon of thoughts and feelings. Perhaps the search for meaning, the mere suggestion that there is something to discover over the next hill, the hope that there is a lollipop for us if we figure the puzzle out, *these* are a part of the very problem. The 'problem' of existence and consciousness itself may be a part of this realm of illusion.

Some might object not only to such suggestions, but to the very nature of this inquiry, fearing its consequences. What if, indeed, we are just pieces of meat on a lonely rock circling a second-rate star near the edge of the Milky Way galaxy. What then? Would it not be better to accept some comforting illusions offered by the mind rather than cutting through our cocoon with a blunt knife constructed from the premise presented above? Much opposition to the study of evolution appears to be rooted in these fears: apparently some believe that the happy illusion of a future life in heaven is better than the fearsome truth of impermanence. If there is no creator, they ask, should we not all just commit suicide? Indeed, our brains may be constructed in such a way as to encourage this very attitude to existence: we may be hard-wired to prefer illusion to reality. Yet it was Jesus, a religious authority and not a scientist, who said that the truth shall set us free.

We can't escape the flow of our thoughts and feelings. Even when we are hiking high in the mountains, with a splashing stream at the side of the trail, we experience illusion. Our minds endlessly replay angry arguments from the past. Our minds speculate what might lie over the ridge tomorrow, or what might happen after the trip is over. Our minds review images that create hope and fear. There is nothing to do but continue on the trail, and appreciate the journey. If we remain focused on the experience of the trail itself, the illusions, at least, are revealed as being illusory.

Even if we are no more than a temporary electro-magnetic illusion on a minor planet, that illusion includes dissatisfaction and pain, as well as pleasure and joy. We can see that others around us are suffering from confusion, craving, disease, old age, and warfare. Indeed, it was that very experience of seeing others suffer that led Siddhartha Gautama to leave his home and begin his own inquiries into the human condition. What, therefore, are we modern people to do? Both Darwin and the Buddha would seem to agree that we have to look at our human circumstances as clearly as we can, however uncomfortable and difficult that inquiry might be. Meanwhile, since each other being around us is similarly caught up in illusion, we can take an attitude of kindness toward them all.

3
The Primate Prison:
The Origin of Self

> In every cry of every Man,
> In every Infant's cry of fear,
> In every voice, in every ban,
> The mind-forg'd manacles I hear.
> (*W. Blake, London, 1794*)

ALTHOUGH HUMANS ARE rather proud of their large brains and their opposable thumbs, it is possible that our crowning evolutionary achievement is the belief in an entity called the self. We are not only conscious, but we have a clear sense of our 'self' as being distinct from the rest of the world. We divide the world into two sets of things: some things are inside our skin (and we label these contents the 'self') and some things are outside the skin (these we call the 'world'). Of the many illusory models that the mind creates to represent the world, from the concept of 'food' to the concept of 'a potential mate', none tops the self-referential model that says, roughly, "Everything within this skin is me. All the rest is 'other'." Psychologists tell us that this view of the world arises early in infancy, when we first begin to understand the difference between ourself and our mother.

Most remarkable, perhaps, is just how powerful and perfect this illusion becomes. It is so real that we rarely question it. Even the brilliant philosopher Descartes began his intense inquiry into consciousness with the assertion that "I think, therefore I am." Thus, our ape ego, has taken that illusion, 'I', and has cunningly inserted itself, not just once, but twice (!), into the most fundamental axiom of Descartes' enquiry.

One of the fundamental teachings of the Buddha was that this 'self' is an illusion. We looked more closely at that illusion in the previous chapter. Gautama's discovery of selflessness was part of his experience of enlightenment, which we set in 528 BCE near the modern city of Bodh Gaya. The self does not exist. Yet the illusion is so powerful that it skillfully perpetuates itself in our lives. Thus it seems it would be useful to say something more about the nature of this illusion.

Others have had the same realization. Shakespeare has Macbeth say

> Out, out, brief candle!
> Life's but a walking shadow, a poor player
> That struts and frets his hour upon the stage
> And then is heard no more. It is a tale
> Told by an idiot, full of sound and fury,
> Signifying nothing. (Act V, Scene v, l. 23-28)

In these few lines Shakespeare tells us about the essential emptiness of ego, and observes that it has the capacity to make us feel that we are real and that we are being watched by an audience. The ego, says the Buddha, is not a thing, not at all, but at best a flame-like process. Yet this process reinforces itself by projecting each of us onto a stage where we imagine ourselves to be acting out our lives. The roar of perception and the rush of emotions like anger inflate this illusory ego and increase the intensity of the experience—yet, in the end it must dissolve into emptiness again.

The reason for the practice of meditation is that it enables us to study this inner phantom that we call the self. A scalpel and dissection manual may be the best tools to begin the study of a human cadaver; in the same way, *shamatha* (calming) meditation and a book such as *The Jewel Ornament of Liberation*, allow us to begin the study of the self. When we sit and relate with the outbreath, we can experience the flame-like process and watch it create the stage and the tale that is so full of sound and fury.

Yet, as we dissect, we find this elegant, painful, and furious tale consists of only a few recurring elements:

> Perceptions: Sights, sounds, tastes, smells, touch.
> Thought: I should do this, she should do that.
> Feelings: Anger, despair, hope, fear, joy.

We discover that existence appears to be little more than a hologram that is generated by an evolutionary projector; yet, all the same, this holographic experience is all-engrossing and seductive. We are like patrons in a special private showing of a film in which we are the lead actor and in which we are not allowed to leave the theatre until our final outbreath and the credits start to roll.

This experience of having a self is, like our other physical and psychological features, an inheritance from generations of ape ancestors, their mammal ancestors, and their reptile ancestors. Only those ancestors who survived to reproduce were able to transmit the projectors to their offspring, and those with projectors and holograms that maximized survival and reproduction predominate today. Whether this tale full of sound and fury is good or bad, whether we should feel happy or sad, is irrelevant. We do it because it improved the survival and reproduction of our primate ancestors. It is what is.

Ego, I, me, myself

Just what is this *I*, this *me*? Somehow, our genes have found a way to cause us to build a mental image of ourselves. This image is something we then consciously seek to protect. *I* must kill the tiger before it kills *me*. *I* must eat because *I* am hungry. *I* am cold. We find ourselves almost unable to utter a sentence or have a thought without reference to this '*I* thing'.

What is it? It is just a model of ourselves, an inner projection, a sensation of structure with which we identify. Yet, it is just another illusion, no more or less real than a cloud or a dream. We dream that there is an *I*, we dream that we are alive, and we are deluded into believing that it is so. All the other experiences—perception, thought, emotion—are deludedly seen to be evidence that this *I*, this *self*, exists as a single solid identity.

Our genes and our infancy trick us into believing that this bundle of experiences exists as a discrete, definable, solid self. And because we all possess this belief (except for some unfortunates whom we label insane), we share this belief, and we reinforce one another's illusions. So great is this illusion, so strong are society's reinforcements of it, that it is doubtful that most readers, first encountering these words, will grasp them. ("What does it mean 'The self is illusory'? Surely this is utter nonsense.")

We are told that the Buddha himself doubted that he would be able to transmit his experience to others. He decided that at least some would understand. He analyzed his experiences, he taught the Four Noble Truths, and he declared that the self was not only an illusion, but the principal cause of human suffering. Experiences, we are told, can be 'self-liberated'. That is, we can still experience perceptions, thoughts, and emotions, but true liberation, enlightenment if you prefer, arises only when these are seen to be natural phenomena in their own right. It is when we put them together into something called a self that the misery begins.

Consider, instead, the Christian tradition. If we puzzle over Genesis and the experience in the Garden of Eden, we may wonder why it was that the tree of the fruit of knowledge was so dangerous. And why did Adam and Eve notice their nakedness and become ashamed after they had eaten this fruit? Perhaps this simple story is an encoded description of the origin of the self. Without the self, we were just another animal, we lived free among them, we did not know we existed, and we felt no shame. Once the self arose, there was no going back. We believed we existed, we could see we were naked, and we felt ashamed of ourselves. This understanding of the story of Genesis is therefore a description of the origin of ego.

Two thousand years later, we still have a society that John Bradshaw, a popular psychologist, calls 'shame-based'. From the evolutionary prospective, shame is just another primate tool for reinforcing the belief that we exist. I feel ashamed, therefore I exist. Shame, by the way, may have a second function: it may make us more controllable by other more dominant members of our tribal unit. But that takes us forward to the topic of hierarchies, how we build them, and then become controlled by them (Chapter 5).

We may also consider that God also arises out of ego, since if we believe in an external deity, the belief strongly confirms that we equally exist as individual ego units. The more we pray to God, the more we convince ourselves that we exist. Just as two mirrors reflect each other back and forth, we say that we are made in God's image (and he made in ours). The mirrors themselves are part of the illusion.

The dance of illusion

Normally, we do not stop to question the experience of being human. As Shakespeare observed, the hologram captures us, and we react. We see images of things we want (food, mates, houses, cars) and the desire to possess these things arises within us and impels us to action. We hardly notice the distinction between the mental image (the picture of the dinner we see with our mind's eye when we are hungry) and the actual experience of the food itself. This confusion is acceptable because we have been compelled to search for the food and consume it. Indeed, we may not even notice ourselves eating because new mental images are arising in our mind's eye even as we eat. What will we do after dinner? We see dancing images that further tantalize us: entertainment perhaps, or socializing, or conflict. How many times have we eaten a meal and hardly tasted it because the image in our mind's eye, part of the hologram projected before us, is more compelling than our actual dinner? Our projections often appeal to us more than our perceptions of reality. "Reality sucks," which I once saw spray-painted on a brick wall while riding a bus, expresses the same view more bluntly. We may even prefer illusion to reality; why else is the television so all-entrancing except that it duplicates a process in our own minds? Indeed, the Buddha himself thought that the capacity of entertainers to create illusion was so powerful that he discouraged his monks from attending theatrical events.

It is not at all easy to see through the theatrical experience of our own ego. This takes us to a fundamental question about the practice of meditation itself. Meditation is difficult. That is one reason why many people avoid it. Even when we have been taught how to meditate by a qualified teacher, we can find countless reasons not to just sit. Why, we might wonder, is meditation so difficult? Perhaps the answer is simple: meditation is difficult because the ego is efficient. Our minds may very well be structured to avoid actions (or inaction) that would dispel the illusion of the self.

Another reason we avoid meditation is that it shows us parts of our self that we have been avoiding. Perhaps Thoreau understood when he wrote, "The mass of men lead lives of quiet desperation." Why should we sit still and feel this desperation, a feeling contemporary writers might call anxiety? The Buddha taught that life is fundamentally unsatisfactory, or, as it is often translated, life is suffering. Compulsive busyness is one means of hiding from this suffering.

There are two reasons for making the inner voyage and encountering the illusion for what it is. The first reason is personal. The Buddha saw thorough the illusions of ego and discovered that it eventually was possible to escape from the misery it creates for us. From this perspective, there is no way to escape suffering except to sit still and confront the system that confuses and traps us, much as if we were a live insect pinned writhing on a specimen card, but afraid to look at the pin. We have to look. As Jesus reminded us, "The truth shall set you free." The problem is that we soon discover the truth about ourselves is annoyingly elusive. Our minds, when we look within, are slippery.

The second reason for making the inner voyage and experiencing the illusion is much larger in scope. We are inspired to look within for the same reason that we are inspired to explore space. It is there, and it is part of our nature as humans to explore it. Further, for those who study the human condition, be they philosophers, evolutionary biologists, criminologists, or literary critics, there is no substitute for actual experience of the ape psyche. Our minds are the very raw material of our own study. It may well be that no one can do truly effective work in any of these fields without first making the trip through one's own inner workings. Otherwise, one is like the literary critic writing on Shakespeare who has never read an original play, or like a doctor who has never dissected a cadaver, or like a tropical biologist who has never been outside her university office.

Exploration of ego

One of the constants of the human condition is our ceaseless mental and physical activity. Call it a kind of compulsive busyness. We all know people who are obviously like this, and sometimes it is easier to see it in others than in ourselves. This ceaseless activity raises clouds of confusion like dust in a desert, or mud in water; it obscures the processes that create the mental experience of being human. Therefore, the Buddha sat still. And, therefore, we too sit still. The Zen master, Suzuki Roshi, says that "What we call 'I' is just a swinging door, which moves when we inhale and when we exhale." The process of sitting still, be it on a cushion, or chair, or under the Bodhi tree itself, is regarded by Buddhists as the essential first step to escaping from our life of quiet desperation. Anyone who has tried this will discover how both our body and our mind rebel: we do not want to sit still, we do not want to experience the flow of thoughts, feelings and impulses that speed through our mental space. We do not want to experience what we are. The beginning student of Buddhist meditation is often troubled to find the degree to which our mind/body stream resists attempts at introspection.

To return to Shakespeare, what does the Doctor tell Macbeth about his wife's misery? Macbeth entreats him

> Canst thou not minister to a mind diseased,
> Pluck from the memory a rotted sorrow,
> Raze out the written troubles of the brain,
> And with some sweet oblivious antidote
> Cleanse the stuff'd bosom of that perilous stuff
> Which weighs upon the heart?

The doctor replies

> Therein the patient
> Must minister to himself. (Act V, Scene III, l. 40-46)

We must minister to ourselves and yet it seems as if the mechanisms for generating illusion have been packaged specifically to resist analysis. And why not? The truth may set us free, but evolution deals with practical consequences of persistence and propagation.

The traditional Buddhist approach is summarized in the Four Noble Truths. Life is suffering. The cause of suffering is desire. There is a way out. The way is meditation. In other words, we sit still so we can see how the suffering and desperation continually create themselves within the mind/body stream. As we practice sitting still and simply experiencing what we are, the restless dissatisfied mind/body experience begins to resolve into parts. Let us look at some of the parts.

Some of these parts we recognize as inputs through perception. Sight, sound, taste, smell, and touch are the five sensory inputs from the world. We live in a five-channel universe. Says Blake "…for that call'd Body is a portion of the Soul discerned by the five Senses" (*The Marriage of Heaven and Hell*). The body and the mind arise as a flicker of sight, then a sound, then back to sight, then perhaps an odour, back to sight, and then a sense of touch. Our world view is assembled from these fragments, yet there seems to be no apparent conscious control of which sense fragment will present itself next. Each sensory input arises separately and distinct from the next. They may appear to flow, but really, they arise one at a time.

Another set of experiences revolves around thoughts. Sitting and relating with our own flow of inner dialogue, we discover that these thoughts seem to come from nowhere, they float around a while, and then dissolve. There are angry thoughts such as "I want to get even with my high school math teacher." There are seductive thoughts such as "How can I connect with such and such a lover?" There are ignorant thoughts "Leave me alone, let me sleep." And then there are elaborate thoughts connected by emotional story lines—while sitting on the cushion we may in a matter of minutes meet someone, propose, marry, have children and divorce, and then discover that we have been sitting still the whole time.

Many of our thoughts are connected to the visual sense. We see in our mind's eye the image of a ham sandwich and discover we are hungry. We see in our mind's eye the image of an old enemy, and we become angry. We see in our mind's eye nothing in particular and become bored with it all. We discover we have a huge movie library, and our mind is playing our stories over and over and over again. For over 2,000 years, Buddhists have watched this kaleidoscopic display and can tell us there are three kinds of stories sorted by the impulse (psychological energy) that carries them along: passion, aggression, and ignorance. Or, desire, anger, and confusion.

At first the flow of perceptions, thoughts, images, and emotions can be overwhelming. With practice, it becomes clear that while our mind appears to provide a constant tidy story told by our thoughts—a single well-edited movie about our lives—what is actually there is more like a cloud of repetitive thought fragments floating in disarray on a sea of churning emotions. It is not a well-edited movie so much as a collection of thought fragments swept up from the cutting room floor. As these thoughts become familiar to us, the raw energy of emotion begins to present itself.

The journey onward, whether led by a Buddhist, a tribal shaman, or a modern Jungian analyst, therefore has constant and repeating elements. This is the inner voyage into the workings of ape mind. And, according to Buddhist traditions, no voyage is more important or more difficult.

The evolutionary biologist, alas, is rarely exposed to these experiences, and so rarely is introduced to the exquisite machinery that produces the ape mind. In the same way, Buddhists rarely have the scientific training to admire the magnificent complexity of the ape mind. Yet, both are describing the same experience.

The barriers to introspection

Why is introspection so difficult? The ego appears to be designed like a sealed black box. Perhaps evolution has sealed it to protect us from being electrocuted or, in other words, to protect us from the raw truth. The effective functioning of the evolutionary process requires us to be convinced by the sound and fury that the stage and the events upon it are real. Today we may spend millions of dollars to create illusions called special effects on theatre stages and in films; the evolutionary machinery of the human mind has had millennia to perfect is own equivalent special effects, Shakespeare's sound and fury. Once we are free (in Jesus' words) or liberated (in the Buddha's words), we are no longer prisoners of the evolutionary machine. We are no longer compelled through life blindly by unconscious agents. We may speculate that the very inability to see that sound and fury signifying nothing has maximized our reproductive success in the past. We are supposed to be driven by illusory wants because the genes that created this worldview have reproduced themselves more successfully than those that did not. The mind machinery is impenetrable because genes that made it so have out-survived and out-reproduced those that did not. The box with the projector within our mind has been slowly welded shut.

Further elements of ego

As we voyage inward, the most obscure and subtle experiences are the most difficult to describe. This is for at least two reasons. First, these experiences are normally hidden from view, and we are trying to describe in words something that is unconscious in us and therefore normally unobservable. Second, even when we can see these experiences, we lack a proper English vocabulary to express them because our language is designed to describe the world of 'outer' experience (thought and perception) rather than the world of 'inner' experience (feelings, emotions, and tendencies).

Yet, it is in emotions and the even more subtle tendencies where we become most aware of our programming. Most of us have experienced anger, and most of us have experienced it too much, so we understand both Buddhist and evolutionary enquiries into human anger and aggression. Indeed, the issue of whether humans (or naked apes) are inherently aggressive has become widespread in popular culture. We have all seen or heard stories on this theme—the opening scenes of the film entitled *2001*, where one ape clubs another with a bone, the film series entitled *The Planet of the Apes*, or the latest pop psychological perspective on gang violence or warfare. We can notice anger because it is such a strong and obvious mental force.

Sexuality is another example of a force that the conscious can see, albeit reluctantly. We are all to varying degrees aware of our centaur-like nature and the sometimes-embarrassing interaction between our conscious impulse to maintain decorum and our unconscious urge to reproduce.

But what of the many other inner forces, those hidden tendencies that are deeply buried far from conscious thought? Those tendencies may be so subtle that we do not feel the slight click as a thought stream shifts from a main line to a sidetrack or back again. The gentle nudge that shifts our awareness from one object in a room to another, that adjusts how far we stand from one another, or where we place our gaze in the elevator, or how we hold our bodies. Such perceptions are processed, instinct evaluates them, and commands are issued, frequently without the least conscious knowledge.

It is here in this hidden realm of the unconscious that we encounter some of the strongest forces of evolutionary biology.

In the 1960s, well before the current generation of sociobiologists and evolutionary psychologists, Robert Ardrey set himself the task of studying social behaviour in animals with a view to understanding humans. He described a wide array of animal studies, with particular emphasis upon other species of primates, seeking the elusive 'tendencies' that program us to behave in certain ways. His list included the following:

- The tendency of young males to form cohorts and travel together;
- The tendency of young males to follow and defer to older males;
- The tendency to adjust our social space, with different tolerances for distances behind us than in front of us;
- The tendency for habits adopted by older males to spread rapidly in a tribe;
- The tendency to form tribal groups led by one or a few older (or alpha) males;
- The tendency for tribe size to be set by roughly the number of separate individuals an individual can recognize;
- The tendency for tribal groups to ignore or attack strangers;
- The tendency of a group to follow its leaders when the leaders act or move.

He also noted that modern society provides a great many stresses on our evolutionary heritage. The family unit of two adults and their offspring, characteristic of twenty-first century humans, is a most unusual structure; most primate societies are organized tribally with short-lived pair bonds and group childrearing. Similarly, today vast numbers of young males compete for very few positions of authority. There can, for example, be only one President of the United States; in tribal society, in contrast, with groups of less than 100 people, many more opportunities arise for males to exercise leadership. Ardrey leaves us with a worrying message that many of the stresses of modern life may be the result of a poor fit between our contemporary social structures and our evolutionary heritage. Alas, Ardrey too seems to have had no exposure to meditation and the inward voyage; one gains the impression that he would have been as fascinated by this voyage as his lovingly described trips to Africa.

In his list of tendencies, there is one missing, perhaps not surprisingly, because this tendency lies beneath all the others and is, according to the Buddhist view, the most troublesome, the most powerful, and the most elusive of all: the belief in a self, the great actor on the evolutionary stage.

A perspective from depth psychology

Ontogeny recapitulates phylogeny. That is to say, evolutionary changes in the physical body that occurred over millions of years are to some extent repeated as an embryo develops into a human. There is, for example, a developmental stage where the human embryo has a fish-like body including gills. Evolution is inherently conservative and, by studying the embryonic development of humans, we can learn about the history of the species, within certain limits of course. Natural selection also acts upon early life history stages, from the sperm and ovum, to the placenta and the embryo, so that even these stages no longer have all the traits of their ancestors. Over time, evolutionary biologists have become equally interested in how ontogeny departs from phylogeny, for there is information not only in the general principles of science, but also in the way in which these rules are modified in response to natural selection.

Erich Neumann, one of Carl Jung's students, applied the principle of ontogeny and phylogeny to the study of human consciousness. He thought he could detect three main stages of consciousness. In the first case, life is unconscious, driven by instincts. As the ego becomes separated from the rest of the mental contents, self-awareness becomes increasingly possible, and the ego increasingly imagines itself to be an individual isolated from the rest of the psyche. Finally, the possibility exists (often later in life) for the ego to begin to re-assimilate the hidden contents of the unconscious from which it initially separated. This third state, he suggests, is central to many spiritual paths. Jung called the latter process individuation, and Neumann called it centroversion. Much of the terminology of depth psychology is unfamiliar to other traditions, but let us look at a few more of Neumann's observations in his own words. He opens *The Origins and History of Consciousness* with the observation that

> The mythological stages in the evolution of consciousness begin with the stage where the ego is contained in the unconscious, and lead up to a situation in which the ego not only becomes aware of its position

and defends it heroically, but also becomes capable of broadening and relativizing its experiences through the changes effected by its own activity. (p. 5)

The emergence of this ego and its self-awareness demands energy. The degree of animation of the ego naturally varies, "ranging from reverie, partial attention and a diffuse wakefulness to partial concentration…, intense concentration and finally moments of general and extreme alertness" (p. 280). When animation is lost, the ego regresses back into the unconscious. Ego experiences this situation as a kind of death.

> The unconscious state is the primary or natural one, and the conscious state the product of an effort that uses up libido. There is in the psyche a force of inertia, and kind of psychic gravitation which tends to fall back into the original unconscious condition. (p. 280)

As the ego begins to emerge, it is associated with a feeling of dread, "an expression of the dawn situation when a small and feeble ego consciousness pits itself against the cosmos" (p. 41). Neumann posits that humans are still in an early age of evolution, for "the organ of consciousness is still at an early stage of development and relatively unstable." (p. 281)

Conclusion

On the trail we are somewhat isolated. It is poor etiquette to talk continuously. On a narrow trail, we may see only the backpack of the person ahead of us. We have the time to note how the intensity of illusion changes from moment-to-moment. A painful memory arises, then the call of a wild bird cuts through that story. Intense feelings give way to the cool expanse of the forest.

When we form large groups, as in cities, the illusions can seem more real. The daily news becomes our reality. At the same time as confused and clouded ape minds multiply and change the world, they tell one another stories about the experience. Each participant believes in its own self. As we multiply, we become an avalanche with the combined psychological momentum of now 7.5 billion, soon even more, confused individuals. Within this mass, thoughts arise and fall away. Emotions erupt and recede. Perceptions are switched on, then off, and then back again. Subtle tendencies adjust our social groups. New social groups form. Masses of humans have the urge to migrate. We continue to make babies even in the midst of famine. We wear expensive watches while other humans starve. Our forests and wild places slowly disappear. And, beneath it all lies a short script, which tells us that we are individuals with a primary responsibility to our own 'I'.

4
Selective Memory:
Maintaining the Illusion

> A human being survives by his ability to forget. Memory
> is always ready to blot out the bad and retain only the good.
> (*V. Shalamov*, Kolyma Tales 1980, *p. 66*)

WE HAVE SEEN how the human mind creates an illusory stage upon which we take our place as actors. This process of creating an illusory world, a kind of cocoon, arises in a precise and well-documented set of steps. The process begins with a sensory input. The mind starts processing this input, and then a critical step occurs, when the input is connected to a pre-existing state that was, up to that point, latent in our own mind. At this precise point, the image is interpreted and becomes part of an illusion. These steps from sensation to interpretation are a kind of psychological chain reaction. Each sensory input is therefore linked with a latent tendency, which itself is based, at least in part, on past experience. Thus, what seems to be happening right now is continually distorted by what has happened in the past. The past is part of the weight we carry. Even on the trail. A certain bird call may remind us of events decades ago. Just how much of our past is it necessary to carry?

The Buddhist tradition has carefully described how the past conditions present experience. For example, in *The Noble Eightfold Path,* the Buddhist teacher, Bhikkhu Bodhi, recounts how each instant of sensory input has the power to activate tendencies that lie dormant in our minds. A simple piece of sensory input, a sight, a sound, a smell, can trigger intense reactions. This happens so quickly that usually we do not notice the small gap between the sensory input and the consequent interpretation. He explains that we can escape from our conditioning, from the cocoon, by remaining aware of this process as it occurs, by taking note of the input before the intense reactions begin. It actually is possible, he explains, to experience this instant of sensation before our mind superimposes a heavy-handed story-line. It is normally a rare occurrence in our lives, but it can be cultivated, particularly through the practice of meditation.

Since our memories provide such a powerful source of intense reactions to the present, we are often advised to let go of what has happened in the past. This advice to let go of the past has a long and rich history in Buddhism. Letting go of the past is one of the critical oral instructions that one great Buddhist practitioner, Tilopa, gave to his student, Naropa, while they were living on the banks of the Ganges River some thousand years ago (Tilopa lived from 988-1069). Tilopa and Naropa could take us on a long digression, but the point here is that this particular single piece of advice is entirely modern. Think of the large number of people who are in psychotherapy sifting through their pasts for the source of their suffering. If it were that easy to let go of the past, the entire industry based upon trauma and recovery from trauma would not be necessary. Instead, the word trauma has become embedded in modern thinking. Indeed, in *The Trauma of Everyday Life*, Mark Epstein has even written an account of the Buddha's own life as a voyage of recovery from the trauma of his mother's death.

It might be easy, therefore, to conclude that memory itself is a source of human suffering. Is that not what Tilopa told Naropa? But here we have a kind of paradox. The past also has vital information for making sense of our lives. Our personal past may indeed have the capacity to distort our day-to-day perceptions of events and to cause suffering. But our collective and cultural past has important lessons that actually clarify our day-to-day perceptions of events. We have learned, for example, that dictatorial governments and warfare are great causes of human suffering. Hence, we try to avoid them when possible. Indeed, we are going to look at Russian memories of the Stalinist labour camps in this chapter. They tell us something important about the human condition, which is why it is important that we do not forget how and why they happened.

More recently, and on the topic of wild nature, we painfully learned that pesticides such as DDT had the capacity to prevent wild birds from reproducing, and we learned that when we limited the use of DDT those wild birds could recover. That is actually good news, which people forget. When I was young, large birds of prey like bald eagles and ospreys were in catastrophic decline. Finally, we passed new laws that prevented the widespread use of DDT, and they were so effective that bald eagles and ospreys are again common. I have seen both in my own forest. Good news. We actually learned an important lesson. Wait a minute, you might ask, how did we get from sensory perception, to memory, to Tilopa, to the Russian gulag, and to bald eagle recovery? Because of memory. Apparently, in spite of what Tilopa told Naropa, memory has an important, even vital role to play in our modern world.

Yes, the past has important lessons. The statement that "Those who cannot learn from history are doomed to repeat it" is not to be taken lightly. In spite of what Tilopa told Naropa, when I fly somewhere, I rather expect the pilot to remember how to use the flight manual, and when I visit my family doctor, I rather expect her to remember at least some of what she learned in medical school. So just what was it that Tilopa was trying to teach Naropa, telling him to let go of remembering? An apparent contradiction like this demands some sort of clarification.

To begin our inquiry, let us look at the role of memory from an evolutionary perspective. This is the biological reality of our human situation. This part is rather simple. In order to make sense of the world, we must be able to label experiences that arise, and we must be able to remember if they are similar to events that have happened before. One could argue that this is the whole purpose of consciousness: we can recognize when current events are like past ones and then take appropriate action. That plant made us sick, that trail led to a dangerous cliff, that predator injured us, and that kind of stone makes a good spear point. If our memories were perfect, then we would know each plant, each bird, each trail, each sight, each sound, and know exactly how to interpret it to maximize our survival. Cliffs are dangerous, water can drown, tigers can bite, snakes can kill ... it is advantageous to remember such things. It is likely that memory arose for just such a reason. But we can't live in a world of memory, and so we must sample our current reality with our senses and continually compare those sensations to stored experience.

The costs of remembering

In spite of the self-evident advantages of memory, it is inescapably true that it is too easy to forget. Let us take a look at forgetting. One could argue that, at some level, it actually seems easier to forget than to remember. That is why each spring I have to re-learn some common birds that come here every year. Why? Perhaps it is just that nature has been unable to build a mechanism to make memory infallible. But it is more likely that there are physical and psychological costs to memory. Home computers remind us that storage capacity has limitations, and added storage incurs added costs. The brain already uses more energy than any other human organ, roughly one-fifth of all the energy we consume. That is a huge cost for humans, particularly during periods of famine. Perhaps energetic costs explain why memory is limited.

But there might be other processes at work here. Might there be other reasons why memory seems to be constrained? We have already seen that there is no reason for our perceptions and minds to tell us the truth. Those organisms with lying minds may survive better and reproduce more than their brethren blessed with clarity. Natural selection operates on only one principle: those who survive and reproduce better than others will become more common.

Perhaps, then, there is an added cost to remembering too much. Life is hard, and memories can be a source of pain. Ask any therapist if you are skeptical. Or, recall the Buddha's statement that life is suffering, the First Noble Truth. Instead, let us ask Shakespeare. Near the very end of *King Lear*, Shakespeare expresses this view of life (through the Earl of Kent)

> Vex not his ghost. O, let him pass! He hates him much
> That would upon the rack of this tough world
> Stretch him out longer (Act V, Scene 3, l. 315-317)

Being alive, says Shakespeare, is like being stretched on the rack. King Lear had his particular misfortunes but even in the United States of America, arguably the wealthiest and most politically-free entity that has ever existed, more than ten people out every hundred thousand are officially recorded as dying by suicide. In other communities, rates of attempted suicide have been quoted at 25 percent of the population. Clearly, some apes find life painful and believe that suicide offers an escape from pain. The strong social opprobrium associated with suicide probably means that official statistics are significantly underestimated, since for the sake of family, many deaths that are suicides may find their way into the category of accidental death (some 35 per hundred thousand).

In *Kolyma Tales*, Varlam Shalamov describes the horror of life in the prison camps established during the Stalinist purges of 1934-1939 in Kolyma, where it is estimated some three million humans were starved and worked to death. His stories predate the epic work by Solzhenitsyn, *The Gulag Archipelago*, which Shalamov was asked to co-author. Says

Solzhenitsyn, "Shalamov's experience in the camps was longer and more bitter than my own, and I respectfully confess that to him and not me was it given to touch those depths of bestiality and despair toward which life in the camps dragged us all." These camps, and Shalamov's tales, describe a stark and cruel world in which the bitterness of life could not be avoided. In one case, a prisoner facing a return to a more severe work regime "hangs himself in a tree fork without even using a rope. I'd never seen that kind of suicide before." (p. 68) Major Pugachov escapes, but as the searching troops close in, he "put the muzzle of the pistol in his mouth and for the last time in his life fired a shot." (p. 103)

This chapter could therefore be subtitled 'the cost of remembering'. Perhaps we all must ride the razor's edge, our brains forcing us to remember to survive, yet if we remember too well, the pain becomes too great to bear. When we examine our own minds, we find, as Shalamov observes in the opening quotation, that important memories fade all too fast. Here we find ourselves beginning the twenty-first century, yet it is all too easy to forget the past few centuries. Who wants to watch those old black-and-white documentaries about the horrors of war, starvation, and extermination? So, we forget. Our minds instead seem easily lulled into dullness by the steady flow of irrelevant ephemera of sports statistics, political theatrics, entertainment, cat photos, and internet postings. Perhaps this is how the ego keeps us distracted from the painful points of reality.

Forgetfulness may be part of the finely-balanced neurological conspiracy that keeps us alive. It may be an empirical consequence of evolutionary biology and the laws of natural selection. Forgetfulness might be a key to building the illusion in which we live. Those who remember too much cannot bear to survive. Those with limited capacity or willingness to remember may beget far more grandchildren than a Philosopher King.

The late twentieth century

My generation grew up in the ruins of the Second Word War. We have already forgotten so much. In the time it took to prepare this manuscript, the generation that fought that war has mostly vanished. Now it is history. Recently, I heard on the radio that three out of every four air crews in bomber command failed to survive the required 30 missions over occupied Europe. The old black-and-white footage can still be seen. Young men in aircraft blew up on take-off and were incinerated in metal cylinders along with their fuel and their bombs. Others were torn apart in the air over Europe by fighters or flak, their aircraft sliced into pieces that tumbled slowly to earth. My father's occupation during this war was repairing the bombers that returned so they could take off again the next day. My father, too, is dead now. His experience has passed on as well.

In my own time there have been the killing fields of Biafra, Cambodia, Yugoslavia, and Rwanda, to name just four. I recite these names here because I, just another confused ape mind, am already forgetting. Shalamov describes how a bulldozer acts like memory.

> The bulldozer scraped up the frozen bodies, thousands of bodies of thousands of skeleton-like corpses. Nothing had decayed: the twisted fingers, the pus-filled toes which were reduced to mere stumps after frostbite, the dry skin scratched bloody and eyes burning with a hungry gleam.... The bulldozer roared past us; on the mirror like blade there was no scratch, not a single spot. (p. 179-180)

"I write poetry," said Allen Ginsberg, "because Russian poets Mayakovsky and Yesenin committed suicide, somebody else has to talk."

Lest we forget

Yet another part of us wants to remember! This is why we have museums, archives, memorials, tombstones, artifacts, pagodas, barrows, and pyramids. In some innate way, we know that we must be continually reminded of those who have gone before us. If we could remember, we would not find our past civilizations producing so many barrows, pyramids and gravestones. Even as the ape ego thrives upon forgetfulness and ignorance, some other part of mind struggles to see the stage clearly. We endeavour to build the memorial that will remind us again and again that there is life beyond our own ego. Shalamov again, on the bulldozer and the mass grave

> How did there come to be such an enormous grave in this area?... I realized that I knew only a small part of that world, a pitifully small part, that twenty kilometers away there might be ... a gold mine with thirty thousand prisoners. Much can be hidden in the folds of the

mountain. And then I remembered the greedy blaze of the fireweed, the furious blossoming of the taiga in summer when it tries to hide in the grass and foliage any deed of man—good or bad. And if I forget, the grass will forget. But permafrost and stone will not forget. (p. 179-180)

The inner propagandist

Our experience is constructed on a foundation of certain biological necessities. One of these is the existence of the self. The self needs to have memories. But memories have costs. We might imagine that the entire industry of psychotherapy exists because of these costs. And, the more that we understand the human psyche, the more we find that memory is very selective indeed.

A great deal has been written about the power of propaganda. By selectively choosing certain facts and ignoring others, it is possible to create realities that trap entire countries. Without propaganda, Hitler and Stalin would have been much less effective. They did not invent anything new; their propagandists were exploiting a mechanism already created by nature. In each of us, the ego, our experience as a self, and our own view of history, are kinds of illusions. We can agree that some of the illusions are closer to factual reality than others, but even those with the best of intentions forget. And, if you can control what an individual forgets, you can control their view of the world and their behaviour in that world.

We may be able to see the efforts of past propagandists and imagine that we are so much more clever. We assure ourselves that we would not have been among the masses wildly cheering a particular dictator, or putting fellow citizens in trains going to death camps. It is harder to see the propagandists at work in our own era. Nevertheless, you may be certain that they are here, in the shadows. Who decides whether we will read about football games or the future of our national park system? Who decides that the death of a particular singer or criminal is more worthy of attention than the death of a Blue Whale or White Rhino? Who decides whether the topic of conversation at tonight's dinner will be the biological basis of climate change or the biological basis of gender, or both, or neither? Political scientists have, not surprisingly, thought a great deal about such matters. In *The Irony of Democracy*, Dye and Zeigler explore the subtle ways in which élites exercise power in human societies. They observe that although élites may have the power to make policy decisions directly, they equally exercise power by dictating what issues will be discussed. "Power is not only deciding the issues but also deciding on what the issues will be." This form of power can be exercised in three ways: (1) élites may act directly to exclude an issue from discussion; (2) subordinates may anticipate the negative reaction of élites and ignore proposals or suggestions that would disturb the élite; or (3) the underlying values of society itself may prevent serious consideration of alternative programs and policies.

Perhaps we seem to have drifted off topic. What has this got to do with memory? The point is that political propaganda is a familiar external illustration of a hidden internal process. Yes, we know (or at least claim to know) that we are all subject to political propaganda. It is part of our perceived external reality. (Of course, the less we notice it, the more effective it is.) My point is that inside us a similar process is occurring, with each unconscious decision to remember or forget. Day by day, week by week our memory is editing the transcript of our own reality. We can't even see the redaction because it is gone. Does this internal propagandist have an agenda? Yes, at least in the long run. What is that agenda? Survival and reproduction.

Pollyanna and Eeyore

The topic of memory has been explored extensively by psychologists. I am going to resist the tendency to write as a scholar and review the field of studies that explore selectivity in memory. But I will offer you my understanding of their conclusions. To start, there is a general consensus that people tend to remember positive events better than negative ones. This has been dubbed the Pollyanna effect, after the character in a children's book by Eleanor Porter who always had an optimistic view of life. One exception to this effect is people who are depressed; they may consistently remember the more negative aspects of their life. I'm not sure that this group has a nickname yet, so let us name them after Eeyore, the pessimistic donkey character in the A.A. Milne series, *Winnie-the-Pooh*. In my experience, neither Pollyannas nor Eeyores are much fun to have around. The very consistency of their ability to edit reality is somehow really annoying. It is just possible we find it so because their sheer tenacity in sticking to their own version of reality challenges our own edited view of reality, the one that keeps us going from day to day. If you get a room full of either personality type at some social gathering, it is definitely time to leave early. If you happen to be an invited teacher at a Buddhist event with such a crowd, you will have your hands full, since each group will have their own take on the First Noble Truth. However, it is more likely that the first group is at home watching a rerun of *The Sound of Music*, while the latter group is at home watching *The Battle of Stalingrad*.

The tendency to edit memories has enormous significance for the legal system, since it turns out that eyewitnesses to crimes can be remarkably unreliable. Sadly, it seems, many people have done prison time, or worse, for crimes they did not commit. Sometimes eyewitnesses just don't get it right—even with basic issues like correctly identifying the perpetrator. People really have been accused and found guilty of crimes they did not commit by eyewitnesses who erred. Memories are not accurate representations of actual events. In a scholarly paper on 'the fallibility of memory', Mark Howe and Lauren Knott explore how witness testimony can be misleading and outright wrong. This is particularly true when testimony is modified by psychological techniques that are used to extract so-called 'repressed memories'. Alas, in some cases, these memories are altogether falsified versions of reality. Apparently, if it is probed enough, the unconscious will create emotionally-charged 'memories' in response to the interviewer.

You may recall that during the 1980s and 1990s there were many children who were alleged to have suffered abuse in Satanic rituals at daycares. One prominent example of this was the McMartin Preschool case, where seven teachers including the owner of the preschool were accused of child abuse and satanic ritual abuse. It became "one of the longest and most expensive trials in California history." Eventually all charges were dropped. Transcripts of the interviews with children show how the 'investigators' created the very memories they were supposedly investigating. More generally speaking, even the familiar technique of using anatomically-correct dolls in court may create sexualized false reports in non-abused children. So, great care is necessary in interpreting other people's memories. The authors describe just how determined some people are to convict others of non-existent acts. Some claim that if you cannot remember sexual abuse from your childhood, that lack of memory is proof in itself that those memories were suppressed. Read that last sentence again. That is the sort of thinking that can take us back centuries to witchcraft trials. The large body of work on fallibility of memory also challenges us to carefully consider modern accusations of sexual abuse, particularly when the memories are not conscious, but extracted years later using questionable methods.

Another body of research shows that people tend to focus upon and remember material that is consistent with what they already believe. If you assign people to read an article on global warming, for example, their prior beliefs on the topic will determine what they remember from the article. Apparently, the raw material from which we build our understanding of the world is heavily edited. According to Margaret Heffernan, this is the result of a rather basic bit of human physiology: when we encounter thoughts that we agree with, we feel the same type of euphoria that a drug addict feels when reunited with their drug of choice. Related studies show that we are better able to find the contradictions in the political speeches of people with whom we disagree than in speeches by people who represent our own views.

By the way, this explains one reason why I constantly refer to dark events like the Second World War and the labour camps of the Gulag. There is good reason to believe that many younger people don't want to remember those aspects of our history, except for a negative comment or two about how Hitler was a bad person. A quick test: how many readers have copies of *The Gulag Archipelago* or *Kolyma Tales* on your book shelves? Now how many have books about Harry Potter or baseball? Come on, admit it. And, it appears to me that when it comes to dark stories, there seems to be a preference for imaginary darkness; contemporary examples include *The Hunger Games* and *Game of Thrones*. The very existence of the word 'game' in the title might be taken as a subconscious reminder that the whole venture is imaginary and thus not too threatening. That is exactly why I re-use these real, historical examples. It is my opinion that they are not only being forgotten, but the forgetting of them allows people to construct false views of reality.

The objective of this book is not to make you feel good, but to explore the biological basis of the Four Noble Truths. To escape, we first have to understand the nature of suffering. Thus, the book was not written to make you happy, but it is just possible that knowing more about Darwin and the Buddha might set you free. Really, if you just want a short period of pleasure, there is always cannabis, alcohol, or sex. Of course, if you look around, you can see where the indulgence in such pleasures has brought us. If, instead, you want to be free, well, that is a rather more daunting task. This is why we are talking about Darwin, the Buddha, and the First Noble Truth at all.

So where does this leave our exploration of ape consciousness? If we forget the past, we are trapped within a view of reality that lacks the essential reference points of history. According to George Santayana (who died the year before I was born), those who cannot remember the past are doomed to repeat it. That, in my mind, is the wisdom behind all manner of physical memorials. If we ignore the past, we blithely imagine that a certain leader will give us an enlightened society, forgetting that Lenin promised more or less the same thing to the people of Russia. But on the other side of the coin, if we remember too much, we may, like King Lear, be driven to madness, or like so many others, driven to suicide. The latest data from the Centre for Disease Control in the United States shows that rates of suicide have increased over recent decades, from 10.5 to 13.0 suicides per 100,000 people. Chögyam Trungpa used to compare life to the experience of sliding down a razor blade. We might interpret his example using the cost of remembering: on one side, too little remembered, and on the other side, too much.

Again some Shakespeare, in this case *Hamlet*.

> To be, or not to be: that is the question.
> Whether 'tis nobler in the mind to suffer
> The slings and arrows of outrageous fortune,
> Or to take arms against a sea of troubles
> And by opposing end them. (Act II, Scene I, l. 56-60)

Yes, it is easy to forget, perhaps too easy. But ultimately, whether we forget or remember, whether we suffer in silence or take arms against a sea of struggles, all of these strategies are but the play of ape ego.

Still, and this is important, just because they are all the play of ape ego and are all illusions in that sense, it does not mean that all beliefs and actions are equally appropriate. Some of my Buddhist friends seem confused on this point. Some seem to think that if you practice meditation, you need not know about history or biology and can pick and choose your beliefs according to mere convenience. Of course a lot of non-Buddhists also pick and choose beliefs according to convenience too, be they Pollyannas, or Eeyores, or just deluded sports fanatics. But at least they are not distorting the teachings of the Buddha to defend this type of behaviour.

Each of us has to live knowing that our view of the world is, and always will be, incomplete. In the same way, our memories of past experience will always be incomplete. It is true that when we sit in meditation, we experience how our memories of the past and our hopes for the future, simply arise and fall away into space. But this should not lead us to assume that all memories are equally unimportant. If we are to live life in a sane way, we need to ensure that our actions are based upon a somewhat realistic illusion of reality. While all experience is an illusion, and it sounds very spiritual to say so, some illusions are closer to biological and historical reality than others, and that realization is also an important part of any spiritual path. Even the Buddha stopped starving himself when he realized that it was counterproductive, and this insight had important consequences for humanity. In the same way, Darwin understood that his new insight into the process of natural selection was deeply troubling to conventional wisdom. This was one reason he delayed publication; but, in the end, like the Buddha, he knew he had to share this new understanding.

It seems that many people in my own community, Buddhist and non-Buddhist, have lost touch with their biological and historical context. Instead, they focus on themselves, their gender, their skin colour, their family history, and the activities of their relatives. And there is often more than a hint of resentment. Ruminations about these topics provide a thick cocoon. Too many seem to be overwhelmed by rather trivial personal memories. I say trivial in the context of human history. My prescription is to contemplate events like the Russian gulag, life during the French Revolution, the effects of the Black Death, or the chaos during the Muslim invasions of Hindustan. Our personal suffering seems intense partly because we lack historical perspective. Historical context is actually an inexpensive kind of therapy. It is also a Buddhist practice, because it helps reduce the incessant impulse to focus on our own thoughts and feelings. Yes, even on the trail we can contemplate history. If our pack feels so heavy that we want to throw it away, we can remind ourselves that during Napoleon's retreat from Moscow, the road was littered with belongings discarded by the demoralized and exhausted troops. In contrast to them, our own pack and our own circumstances are quite bearable. And we have summer weather while they were dying of exposure.

Memory is imperfect. And, it is good to let go of memories when we are holding on to personal resentments. At the same time, human history is larger than any one person, and human history needs to stay alive in our minds. A good book on history actually can reduce the intensity of our personal cocoon. There is a lot more going on in the world than our own family drama. A good book on history also has a second value. It helps us build a more accurate view of the human condition and our place within it. In order to practice Right Conduct in our lives, we need to remember the lessons our ancestors learned the hard way.

5

The Urge to Impress:
Priests, Kings, and Dominance Hierarchies

> A *linear dominance order* with very few reversals or
> triangular relationships characterised each troop, the males
> generally dominating the females.
> (*Brown 1975, p. 254, italics his*)

ON A HIKING TRIP, some days are harder than others. Today we have some hills to climb, and some switchbacks to negotiate too. Along the way, we are going to chat about another pattern found in humans: our tendency to organize ourselves into social hierarchies. We are again looking for common patterns that occur in nature and which our companions, Darwin and the Buddha, might wish to discuss. We have certainly had a good many conversations about craving. It is time to change the subject. And, for a diversion, let us start by describing a large and old compendium of human behaviour: *The Golden Bough*, by Sir J. G. Frazer. This is an enormous compilation of human religious activity. If we can find some common themes, we can reasonably conclude that we have discovered a pattern (or what Jung called an archetype) worthy of conversation. *The Golden Bough* (at 12 volumes!) is too heavy to carry along on our hike, so we'll begin with a short summary.

The King of the Wood and *The Golden Bough*

The Golden Bough is a compilation of beliefs and practices involving magic and religion among peoples throughout the world. According to the biographical note (p. ix) that accompanies my abridged version published in 1922, "Between 1905 and 1915, it increased until it extended through twelve volumes. It had become the largest report of ancient magical and religious practices ever compiled by a single anthropologist." In his preface, Frazer explains that his primary aim is

> ... to explain the remarkable rule which regulated the succession to the priesthood of Diana at Aricia. When I first set myself to solve the problem more than thirty years ago, I thought that the solution could be propounded very briefly, but I soon found that it was necessary to discuss certain more general questions, some of which had hardly been broached before. (p. xiii)

The little woodland lake of Nemi lies in the Alban hills of Italy. On the northern shore of the lake stood the sacred grove and sanctuary of Diana of the Wood. In the woods, begins Frazer, lurked a single man.

> In his hand he carried a drawn sword, and he kept peering warily about him as if at every instant he expected to be set upon by an enemy. He was a priest and murderer; and the man for whom he looked was sooner or later to murder him and hold the priesthood in his stead. Such was the rule of the sanctuary. A candidate for the priesthood could only succeed to office by slaying the priest, and having slain him, he retained office till he was himself slain by a stronger or a craftier. (p. 1)

Further, says Sir Frazer, "It is no longer possible to regard the rule of succession of the priesthood of Diana at Aricia as exceptional; it clearly exemplifies a widespread institution." To understand his conclusion, it is first important to know that kings and priests were interchangeable in many societies. The profession of magic, according to Frazer,

> ... shifts the balance of power from the many to the one; it substituted a monarchy for democracy, or rather an oligarchy of old men; for in general the savage community is ruled, not by the whole body of adult males, but by a council of elders. (p. 31)

In such as case, "shrewd rogues" are most likely to attain the height of ambition, that is, dominance over the rest of their neighbours, and such men, according to Frazer, included both Caesar and Augustus. Indeed, history is full of rulers who were killed by subordinates who then in turn ruled: examples range from ancient Rome to the most recent tyrannies of Africa. Democracy, in this context, can be thought of as an institutionalized means to depose an elder male without the necessity of killing him.

One way to illustrate the importance of hierarchy to humans is to look at our language. We have a great many words to describe a dominant male ruling over subjects. It would appear that this is something that humans have talked about a great deal.

Hierarchies in ape communities tell us something about ourselves

Frazer was primarily concerned with magic and religion, which from our perspective, are social customs that arise out of archetypes created by selfish genes seeking to ensure their own propagation. What might lead to the custom of rulers being killed and replaced by younger contenders for the throne or priesthood? Of course, Frazer is describing the situation in tribes of primates.

The competition within species for access to mates is intense. Brown has summarized the role of aggression in primate social organization as follows

> ... primarily includes acts of attack, escape, threat, defence, and appeasement.... It is convenient to group these behaviours together because (1) they are functionally related to intraspecific, competitive situations and, consequently, to the dispersion pattern of the species, (2) they are intricately related motivationally and physiologically, and (3) they tend to occur together in time and space. (p. 40)

In tribes of animals there is a balance between aggression driving individuals apart and cohesive behaviour bringing them together. Cohesion is maintained primarily through family units, with offspring tending to remain with parents and males being attracted to females. The chief factor opposing cohesion is aggressiveness, and within the group, this creates a dominance hierarchy. Dominance relationships are most conspicuous in multi-male troops. Consider the vervet monkey in Kenya, where troops had a mean size of 24 individuals with an adult sex ratio of 2.1 females

RULER

- boss, CEO, chairman, directore, manager, president
- corporal, general, lieutenant, major, marshal, officer
- caesar, caliph, emperor, king, master, raja, seigneur, sovereign
- autocrat, despot, dictator, governor, taskmaster, tyrant
- chief, head man, leader, mayor
- alpha male, male dominant, partriarch
- elder, guide, guru, pilot, priest, rabbi

attendance, capitulation, compliance, cringing, deference, devotion, grovelling, homage, humility, meekness, obedience, prostration, service, subjugation, submission, surrender, yielding

The dominant male archetype is well-represented in the human vocabulary. Here is a partial list of the names of authority figures and some of the emotive states associated with obedience in the English language. Our literature and our films continually replay simulations of life with various kinds of rulers.

per male. "A *linear dominance order* with very few reversals or triangular relationships characterized each troop, the males generally dominating the females" (Brown 1975, p. 254, italics his). Dominant males achieved the most copulations, the first-ranking male achieving twenty-six during the study, the second ranking male three. After surveying similar studies, Brown concludes that dominance relationships in multi-male groups are closely connected with frequency of copulation.

Much of the early work on primate dominance hierarchies tended to emphasize the male perspective; females may be faced with a different set of conflicts. Females may prefer high-ranking males, but may not always have access to them. Further, "[n]o one is sure what females really gain by mating with high-ranking males rather than with their subordinate troop mates, and females *always* turn to low rankers when high rankers are unavailable." (Small 1993, p. 160, italics in original). The traits which females prefer in males and the strategies they employ to ensure matings with such males have "riveted researchers in animal behavior since the early 1980s" (p. 99). Possible traits females might consider include (1) social rank, (2) health, (3) intelligence, (4) familiarity, (5) novelty, and (6) paternal ability. Consider the first then last in this list.

High social rank of a male provides a female with assurance of plentiful resources for nurturing her young. "She is a well-off woman that is a rich man's wife," says Ailill to his wife, Queen Medb, in the Celtic epic *The Cattle-Raid of Cooley*. This same urge to increase the status of her husband apparently drives Lady Macbeth to pressure him to assassinate Duncan, the visiting king. Using both seduction and shame, she appeals to his male vanity.

It is possible that females consciously seek males who demonstrate good parenting abilities. Studies of vervet monkeys using one-way glass have shown that males are much more friendly to the offspring of females when they are aware that the mother is watching. When females met males who had been unkind to their offspring, they threatened and attacked them. Thus, says Small, females are not only attentive to the parenting skills of males, but males will alter their behaviour to impress females.

Returning to the male perspective, the only way in which an individual can reproduce is to locate a female. But there is an essential asymmetry: males could, in theory, produce many thousands of offspring, whereas females are limited by the number of young they can carry through gestation, normally a score or so of young at most. This asymmetry of potential reproductive outputs means that males are in competition for females. This leads to a competition in which males are selected for their ability to win out over other males.

Sexual dimorphism is thought to arise out of this competition. Male gorillas and orangutans are twice the size of the females, armed with large canine teeth, and clearly labelled with characteristics such as silver backs. At the same time, however, male gorillas have relatively small testicles for their body size. Because the dominant male has uncontested access to a harem of females, only a relatively small amount of sperm is needed to ensure reproductive success. If this explains why male gorillas are large, aggressive animals with small testicles, how does such logic apply to other primates? Chimpanzees are minimally sexually dimorphic in comparison with gorillas, but their testicles are proportionately larger. It appears that in chimp societies, dominance hierarchies are much more complex; Miele suggests that, in such cases, reproductive success requires more of a Machiavelli than a Mike Tyson. Since dominant males do not have exclusive access to females, their sperm must compete with those of other males. In promiscuous species, the logic goes, the male that injects the most semen has the best chance of fertilizing the egg; thus large testicle size provides a competitive advantage. Humans are in between chimps and gorillas in both sexual dimorphism and relative size of testicles. According to what we know of ape societies, humans also appear to be intermediate in terms of promiscuity and polyandry.

In any case, social hierarchies and sexual behaviour are intertwined. The traits associated with competitive success in apes therefore vary with the definition of success and the social system. In gorillas, large size and weaponry allow males to competitively exclude other males from reproduction. In chimps, social dominance is less clear-cut; cunning and testicle size appear to be the traits that maximize reproductive success. It is no wonder that humans find male-female relationships to be difficult, even troubling. Perhaps if our fellow humans knew more about our near relatives, including chimpanzees, bonobos, and gorillas, we might better appreciate the constraints of our own human nature and perhaps have a little more sympathy for one another.

Of course, we can look further into the animal kingdom than just our fellow apes. Darwin, for example, was an expert on barnacles and wrote an enormous monograph (four volumes long, again too heavy to take on a hiking trip) that deals with both fossil and living species. In some species of barnacles, he found the male had no mouth or eyes and was reduced to little more than a sperm-producing entity that lived inside the reproductive tract of the much larger female. Sometimes multiple males lived within a single female. Sometimes the larger animal was not a female but a hermaphrodite. We may reasonably assume that their emotional lives are therefore different from ours, although, since they do not post on the internet, we can only guess what male and female barnacles have to say about one another.

Status and dominance in ancient Rome

Now for some historical examples. In human societies, one dominant individual can extract resources or mates from the regions that they control. It is instructive to re-read historical texts as a treatment of three recurring themes: the seizing and holding of power by force, the extraction and diversion of resources to the dominant and his supporters, and an excess of reproductive effort by the dominant. Instead of *The Golden Bough*, let us consider just a few cases from the well-documented history of Rome.

Sulla (138-78 BCE) used a military force and massacre to overthrow the democratic leaders of Rome. According to the Greek historian, Plutarch (c. 46 - c. 120), Sulla was a "profuse debauchee," lived in "lewdness and adultery," and "impoverished and drained the city of her treasures ... although he daily gave up the wealthiest and the greatest families to public sale and confiscation." (Plutarch, p. 388)

When he captured Athens, blood flowed from the market place through the gates and into the suburbs. When he seized power in Rome, Plutarch says he hired mercenaries to kill thousands of his opponents audibly even as he spoke in the senate. "This gave the most stupid of the Romans to understand that they had merely exchanged, not escaped, tyranny" (Plutarch, p. 384).

Within a decade of the death of Sulla, Durant (1944) describes how the Romans

> ... milked the provinces at a rate that would have angered their predecessors with envy.... When Caesar went to Farther Spain as proconsul in 61 he owed $7,500,000; when he returned in 60 he cleared off these debts in one stroke. Cicero thought himself a painfully honest man; he made only $110,000 in his year as governor of Cilicia and filled his letters with wonder at his own moderation. The generals who

> conquered the provinces were the first to profit from them.... Pompey brought in from the same region [the East] $11,200,000 for the Treasury and $21,000,000 for himself and his friends; Caesar took literally untold millions from Gaul. (p. 129)

Further,

> If any wealth still remained, a flock of entrepreneurs came in from Italy, Syria, and Greece, with Senatorial contracts for "developing" the mineral, timber, or other resources of the province... (p. 130)

The turnover of leaders was often rapid. "In the thirty-five years between Alexander Severus and Aurelian, thirty-seven men were proclaimed emperors." (Durant 1944, p. 628)

Let us briefly consider five examples from the first few centuries of the first millennium: Caligula (12-41) encouraged citizens to name him in their wills, taxed everything including the earnings of prostitutes ("as much as each received for one embrace"), and "had rich men accused of treason and condemned to death as an aid to the Treasury." (Durant 1944, p. 267) Further, "... there was scarcely one women of rank whom he did not approach" (p. 266). He was killed by a tribune of the Praetorian Guard who was apparently disgusted in part by being routinely given obscenities for passwords. Nero (37-68), like Caligula, says Durant (1944) "... exhausted the treasury with his extravagance..., confiscated the wealth of prominent citizens, stripped temples of their offerings and melted down the silver and gold statues they contained" (p. 279), and lived amidst a stream of sexual licence. Eventually revolt led to his condemnation as a public enemy, and, fearing death "after the ancient manner," he killed himself. Commodus (161-192) had an advisor, Cleander, of whom Gibbon says "Avarice was the reigning passion of his soul, and the great principle of his administration." In the space of three years he accumulated "more wealth than had ever yet been possessed by any freedman" (Gibbon 1776, p. 37). Another of his ministers, Perennis, "obtained his post by the murder of his predecessor.... By acts of extortion, and the forfeited estates of the nobles sacrificed to his avarice, he had accumulated an immense treasure" (p. 36). The reign of Commodus deteriorated into one of the worst chapters of Roman imperial despotism, and eventually his mistress arranged to have him strangled. Of Maximinus (235-238), Gibbon writes "... the whole mass of wealth was at once confiscated for the use of the Imperial Treasury. The temples were stripped of their most valuable offerings of gold and silver, and the statues ... were melted down and coined into money" (p. 71). When there was an uprising led by the Senate, he attacked with his army but finally, abandoned by his guards, was slain in his tent along with his son. Elagabalus (204-222), with "capricious prodigality ... lavished away the treasures of his people in the wildest extravagance" and had a "long train of concubines, and a rapid succession of wives." He was murdered by the Praetorians and his mutilated corpse dragged through the streets of the city and thrown into the Tiber (Plutarch, p. 60-61).

The evidence for these resources being diverted to reproduction is qualitative rather than quantitative, since historians seem to treat the subject with either moral indignation (e.g., vice, infamy, lewdness, adultery, wickedness, passion) or undue discretion (e.g., "The unbounded licence of indulging his sensual opportunities" (Plutarch, p. 38)) rather than giving actual estimates of numbers of mates and offspring sired. In one slight exception, Gibbon tells us that Maximinus had twenty-two "acknowledged concubines" (p. 72).

Status and dominance in an aboriginal culture

Now let us look at a different example, an aboriginal society on the west coast of North America. The Keatley Creek site near Lillooet, British Columbia, is one of the largest aboriginal villages discovered by archeologists in western Canada. There are 115 depressions from semi-subterranean houses, and representative examples of different sizes have been excavated. Different house sizes are, of course, associated with status differences in a community and give an indication of inequality and hierarchy. The archeologists begin with some observations on human behaviour.

> ... A number of powerful forces ... militate against the formation of large multifamily residential groups. These forces include less-efficient heating, different control over stored surpluses, and geometrically increasing possibilities for conflict and fissioning within the group. In order to overcome these centrifugal tendencies, a hierarchy possessing some real degree of power must exist

> to arbitrate disputes and allocate group resources. The larger the ... group, the more hierarchical and powerful we expect its administration to be. (Hayden and Ryder 1991, p. 53)

Fish were a primary source of protein in this society, and there was private ownership of the best fishing rocks, which were passed along though families. The largest families owned slaves and owned houses that were veritable mansions in comparison to others. When excavated, they contained large numbers of items associated with wealth (e.g., soapstone pipes, dentalium shells, elk-tooth pendants) along with a far higher number of tools. The smaller housepits contained exclusively pink salmon vertebrae, the easiest species to catch and least desirable; larger housepits included the more desirable sockeye and spring salmon bones.

> Competitiveness, individualist ethics and hierarchical ranking of individuals and families were engendered by advantages gained from controlling surpluses. It certainly appears that residents of this large housepit had greater access to important economic resources.... (Hayden and Spafford, p. 133)

Hierarchies and reproduction: what we learn from Genghis Khan

There is one more important issue about hierarchies and human behaviour. It would seem likely that one of the principal reasons that hierarchies exist is to multiply the genes of the individuals at the top. That is why sultans had harems. That is why victorious armies often killed conquered men or sent them to the mines, while taking the captured women home as concubines. Consider the example of Genghis Khan, who created one of the largest empires in history. The Mongol Empire stretched from eastern Europe, including modern-day Russia, to the Pacific Ocean. Even Poland, Hungary, and Austria were successfully invaded. It now appears that Genghis Khan left a great many descendants. A scientific paper titled "The Genetic Legacy of the Mongols" reports that one Y chromosome lineage that originated in Mongolia, is now found in human populations across Asia, roughly within the boundaries of the old Mongol Empire. Eight percent of the males in large parts of Asia, about 16 million men, have this Y chromosome. As the paper puts it delicately "he and his close male relatives had many children." Part of the story might be that women were regularly taken as slaves by victorious armies, and as Khan, he would have had his pick of the captive women. Part of the story may be that his power allowed him to put his male relatives and his children into positions of power. It is probably some combination of the two. So, it appears that being at the top has its evolutionary rewards. More recent studies have identified other widespread Y chromosome lineages that could have been left by other warlords. Social prestige has clear consequences for evolution.

So, looking across the animal kingdom as a whole, it is plausible that when a male reaches the top of a hierarchy, his brain begins to tell him that time is short, and it is time to make as many offspring as possible before he is replaced by another male who is going to do the same. The next thing you know, instead of doing his job, he is having sex with an intern on the presidential desk, or something like that. It is possible that a ruler might climb to the top of the hierarchy with the best of intentions, but when he gets to the top, the old ape (or stallion, or lobster) physiology takes over and floods the brain with messages to make lots of babies now. We certainly know that this is how many animal hierarchies have worked for millions of years. We now have some further physiological details to add: the leader's brain may be so flooded with a cocktail of neurotransmitters that he thinks he is on top of the world. Perhaps this even explains in part why so many rulers seem to lose touch with reality and fail to meet our expectations. Perhaps they really cannot see that there are problems because they are on top. Like all of us, then, the leader is trapped within a dream, it's just a different dream than the rest of us have.

Of course, I can already hear the suggestion that if we just put women at the top of the hierarchy, it would all be different. Here again, evolution provides a message. Matriarchies are just not that common in wild nature, and there is likely a biological explanation. The best way for a woman at the top of the hierarchy to multiply her genes is to put a son on the throne. After all, even if she were to start young, she could leave at best some twenty offspring in her lifetime, while a son could produce many more than that in just one year. History is full of examples of women who schemed, even against their own husbands, to make their son an emperor. But since this is a book on Darwin and the Buddha, we will have to stop here with this history. The history after all, is here for one purpose, to illustrate human behaviour. Genghis Khan's Y chromosome lineage speaks for itself.

We have inherited dominance hierarchies from invertebrate ancestors

Those parts of our brain, and the neurophysiologic processes that underlie dominance in humans, are extremely ancient indeed. According to Jordan Peterson, they can be traced back in evolutionary time to bottom-dwelling, marine invertebrate ancestors that existed even before life emerged onto land. Even lobsters have a serotonin-based neurophysiology; they too struggle for status and become depressed when they lose. Lobsters are, of course, just one selected example of marine invertebrates as a whole. But such creatures share a common set of ancestors with us. And, bottom-dwelling marine organisms like lobsters are territorial creatures, just like us. (Or, perhaps I should say, we are territorial creatures, just like them.) The main resource for lobsters is a constant rain of debris from higher levels in the ocean. The quality of a territory is determined in part by the rate at which debris falls, and then by other qualities such as the availability of shelter. The most successful lobsters have a high-quality territory. Lobsters fight one another for access to such high-quality territories. Lobsters also have an inherited set of traits that are associated with fighting ability. Peterson describes how, when two males encounter one another, they dance around like human boxers, raising their claws, and gauging their abilities against those of their opponent.

The lobsters that win these encounters seize the best territories. These dominant lobsters are the ones who attract females and who leave the preponderance of young. These lobsters also have high levels of serotonin and low levels of other chemicals like octopamine. This is the neurochemistry of a winner. Lobsters that repeatedly lose fights become depressed and have lower levels of serotonin.

You can read more about this work in Peterson's book, *The 12 Rules of Life* (conveniently in Chapter 1, along with a discussion of how this affects treatment for depression). The point here is that the very neurochemistry of dominance in humans precedes not only the origin of primates but also of mammals. It is instructive, therefore, to watch the battles for dominance within a democratic political party, or within a cruel dictatorship, and realize that we are actually watching an evolutionary behaviour that we share with bottom-dwelling marine ancestors. It is so ancient that we can think of it as a small, independent sub-routine capable of running without any input from the intellect at all. Indeed, to turn it all around, we might very well argue that the presence of the human intellect is a consequence of the serotonin-based dominance hierarchy, since intelligence may have arisen as a tool for winning conflicts within our own species. Perhaps for humans, intelligence is our version of big front claws.

Do you feel good about yourself today? Peterson suggests that part of our evolutionary heritage is also a 'primordial calculator' that continually monitors where each of us is positioned within our own society. If this calculator judges you are near the top of the hierarchy, serotonin flows abundantly and you feel confident and calm. If the calculator judges that instead you are near the bottom, serotonin flow is restricted. You feel a sense of desperation and grab impulsively at any opportunity for resources or mates. This imagery may have a good deal of value in understanding the cult of victimhood. By definition, only a few people can be at the top of a dominance hierarchy, and hence, the majority of individuals will be lower in position. Thus a lot of people are going to feel badly about themselves. And, in the right circumstances, they will not only feel badly about themselves, but feel empowered to commit impulsive acts of violence when they are given the opportunity.

Returning to history, when the mobs stormed the Bastille on 14 July 1789, initiating the French Revolution, it is evident from witnesses that they were being driven by anger and resentment as well as hunger. Who wants to be at the bottom of the hierarchy? And, in close succession, the rulers went from a king (Louis XIV), through a variety of revolutionary committees, to The Directory, and eventually to Emperor Napoleon. Over a decade, thousands of deaths, and immense chaos, the people at the top of the hierarchy changed. And, at the bottom of the hierarchy were mobs, particularly the *sans culottes*, who aroused nearly as much fear in the revolutionary Directorate as they did in the king. This is history in all its bloody reality. The same story could be told about the Russian Revolution (Lenin temporarily on top) and the Chinese Revolution (Mao temporarily on top). And, from the perspective of evolutionary physiology, it is just a continual dance of who will be the dominant, with the most serotonin and the best food and mates. A hierarchy necessitates a lot more people at the bottom, which may help explain what Jesus meant when he said, "The poor shall always be with you."

In conclusion, status within many animal communities confers preferential access to resources and to females. It therefore confers two important, indeed essential, evolutionary advantages. Craving for status may therefore be a primordial urge within the ape mind. You might even say that this book is the product of my own need for status, demonstrating the very general principle it describes.

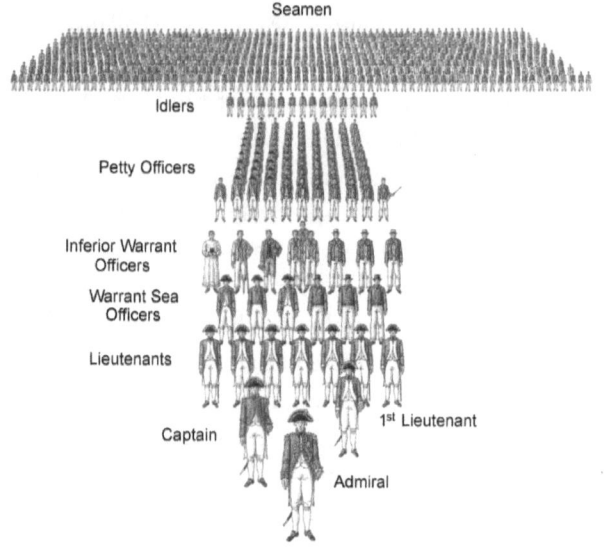

The world is full of hierarchies. This historical example from a children's book shows the hierarchy on a typical British man-of-war during the Napoleonic Era. Many kinds of wild animals also live within hierarchies. Even plant communities are frequently dominated by a small number of large individuals.

Darwin and the Buddha and their hierarchies

Once you start looking for them, it becomes apparent that we live our lives nested in multiple hierarchies. Darwin lived in a country structured by a monarchy (much of his life ruled by Queen Victoria) and by the Church of England (under the Archbishop of Canterbury). The Royal Society, where his scientific account of evolution was first presented, was also a hierarchy. Darwin's famous voyage on the *Beagle*, under Captain Fitzroy, was also time lived within a strict hierarchy. Indeed, in my lectures I have used the British navy as a human example of hierarchies that arise in all manner of ecological communities, including coral reefs and forests. On any ship, the captain, of course, gets the biggest room and the best food.

The Buddha too was born into a hierarchy, with his own father, Raja Suddhodana, subordinate to a much more powerful regional leader, the King of Kosala. According to *The Historical Buddha*, the military hierarchy was independent of this civil hierarchy, with local warriors reporting not to the Raja, but directly to the King of Kosala, to help ensure that local rulers did not have a convenient army with which to rebel. The Buddha walked away from this hierarchy and at one point lived alone in the forest. Yet over time, after his enlightenment, a new hierarchy grew around him. His disciples were ranked by the length of time they had been his students and by the degree of attainment and understanding of his teachings they displayed. As we shall see in the final chapter, other males even challenged him for the right to lead the community. (On another occasion, a woman accused the Buddha of fathering a child with her, an event that resonates with modern concerns about how to judge accusations made against leaders in hierarchies.) Most Buddhist organizations today are themselves hierarchies with single leaders; one example is the Karmapa, who leads one of the larger Tibetan Buddhist lineages, the Karma Kagyu. The fact that there are currently two contenders for this position shows that Buddhist organizations are not immune to the tensions created by hierarchies.

If there is a general lesson, it is a simple statement of reality: this form of social organization is widespread and seems to be deeply rooted in human nature. Since the same sort of organization occurs in many other kinds of animals, including horses on windswept prairies and lobsters deep in the ocean, we could say that this is deeply rooted in animal nature. Taking the point one step further, the same thing happens in forests: in any piece of landscape, most of the water, light, and nutrients flow to a few dominant trees. They take most of the resources, and they produce most of the offspring. So, it would appear that dominance hierarchies are inherent in wild nature in general. All of which is not to say that these are inherently good, or inherently bad, but rather, that the pattern does offer a simple description of what *is*.

If there is a political lesson, it is that when you replace one leader, another naturally arises. Classic example: Lenin promised freedom from the Czar and then gave us Stalin. All too often, people raging against a particular hierarchy are hoping, perhaps consciously, perhaps not, that they will be higher up in the replacement hierarchy. We will return to this topic toward the end of our hiking adventure when we consider possible implications for human governance. In any case, we are certainly trapped within a widespread and powerful pattern in human nature, one that appears in both secular and spiritual guises. We can also see attempts in human history to limit the power exercised by the dominant individual: parliaments and elections are two obvious examples.

Hierarches and female choice

We have already covered a lot of ground, and it seems like it is time to call it a day and cook some dinner. A hiking trip takes some careful planning—we need some adventure, and we need to be realistic about what our companions can handle in one day. I suspect that some of the people who have joined in our conversation might feel that this chapter does not have enough to say about the female half of the human species. So, we are going to continue our conversation for just a little longer, but take a seat on these rocks. We have reached an important conclusion: hierarchies are a fact of nature. And males who reach the top have excellent access to resources and many opportunities for reproduction. What does this mean for us as individuals?

I fear that at least one group of readers may have decided already that our conversation depicts nothing more than oppressive patriarchy, just another example of how unfair the world is to women. Let us look at this proposition more carefully.

We should perhaps begin by noting that hierarchies are also painful and exploitive to the vast majority of males. The lobster king has the best territory and access to many females. The elephant king rules his herd and has access to many females. The human king has a fine palace and access to many females. However, these individuals are the exceptions, the one percent. The vast majority of males survive near the bottom of the hierarchy. In human societies they are the field workers, the coal miners, the assembly line hands, the janitors, and the construction labourers. And, yes, the cannon fodder too. They toil their entire lives with little respect and die in relative obscurity, often from work-related ailments ranging from black lung disease to repetitive stress injuries to bullet holes and to PTSD. Merle Travis wrote a song about a male at the bottom of a hierarchy, a coal miner: "You load sixteen tons, and what do you get? Another day older and deeper in debt."

There is a line of argument, possibly a strong argument, that females reap important benefits from this type of social organization. When men are pitted against one another, it sorts them, and allows women to choose the most desirable mate. In indigenous American tribes, the sorting may have been by knife fights or hunting contests; in modern times, the sorting may be by education or technical prowess or yacht size, but in each case, the men are conveniently arranged in a hierarchy, allowing informed female choice about their relative abilities. Nearly all sports fall into this category too: they are public displays of relative athletic prowess. There is a good deal of information about how women react: to put it simply, women generally prefer the winners. The general term for this is hypergamy: once men are arranged in a hierarchy, a woman can identify the most desirable mates and seek to maximize her own status through mate choice. A more general term for this phenomenon is 'asymmetries in partnering preferences'. Yes, females are making choices.

Female choices, you ask? Surely not, we are told that women are oppressed and choiceless and therefore angry. It is not quite that simple. A great deal has been written about female choices. Here we want to focus on one particularly important female choice: whether to mate and, if so, with whom? There is a book by this very title: *Female Choices*. Meredith Small is an anthropologist, and it is worth repeating her observation: "Males and females are caught in conflict because first and foremost they need each other to make babies and pass on genes, but they use radically different strategies to attain that goal." Males in hierarchies offer females information and choice.

Let us address this topic with an explicit and painful point. Many religious communities, including Buddhist ones, have been damaged when high status teachers (mostly men) have sexual encounters with their students (mostly women). The typical narrative is this: the male in the position of authority is exploiting the vulnerable woman owing to his higher social privilege. There is, however, a complementary narrative: the other

Sexual selection is a powerful force in nature. Some, perhaps many, of the characteristics of males are determined by female choice. A classic example is the peacock's tail, but there are many more examples in wild nature. Here is another example, a page from The Descent of Man, showing a species of hummingbird called the white-footed racket tail, female (left) and male (right).

half of the equation may be that women are exhibiting hypergamy and are attracted to men higher in the hierarchy as their preferred sexual partners. Yes, even at a weekend spiritual retreat, the men are organized by status, and it is clear to everyone who is near or at the top. They often arrive in their own private vehicle and with their own entourage. Just pay attention and you will see for yourself how efficiently males are sorted. Who is staying in a tent, who has a single room at the retreat center, and who is staying in a local mansion? The unhappy situation of sexual misconduct may have more than one narrative. Both narratives may be simultaneously true. It is just possible that participants of each gender are acting out impulses that arise out of roles deeply buried in human evolutionary psychology. This does not make such behavior right, nor does it explain it away, but it might very well explain just why these socially disruptive actions emerge repetitively in spite of our best intentions.

Let us move to a less contentious example of mate choice and sexual selection. Darwin himself wrote at length about sexual selection in a wide array of animals. Sexual selection, in general, refers to the ability of females to select for males that meet certain criteria. (The reverse too can occur, but here we are focussed upon female choices.) The classic example is the peacock's tail. It does nothing for the male, and indeed it has costs: it takes resources to construct and maintain that tail, and it likely makes the male more vulnerable to predators. So why does the peacock have a huge showy tail? It is because past generations of females have chosen to mate with those males having the most showy tails. Yes, from one point of view, that enormous and elaborate peacock has been produced by female choice. Darwin gave many more examples and illustrated some of them with line drawings in several chapters of his book *The Descent of Man*. More than a hundred years of subsequent scientific research have extended this work enormously. Here is the basic conclusion: female choice is a powerful force in nature. Hence, it would appear that some of the shape, size, and behaviour you see in males is the result of mating choices made by past generations of females. Men, then, are in some ways the kind of men that their great great great grandmothers chose as mates. Other men were considered so undesirable that they didn't find mates at all, and they have genetically disappeared.

Now, since I am tired of recounting evolutionary biology and reading about sexual selection in birds (and lobsters and barnacles), and since I am eager to call it a day and have a drink from my canteen, I will wrap up this excursion with a couple of personal anecdotes about female choice.

In a memorable conversation with a woman who was a professional matchmaker, I was informed that all the men on her client list had to meet two criteria: they had to be tall, and they had to have a full head of hair. The need for them to be financially secure was obvious. At first I thought she was making a joke, but she was in fact quite serious, and she said she would not even consider men who did not fulfill those two primary physical criteria. There was simply no market for them. Later, thinking that I might have had got it wrong and still worried about my single friend, I called her back to double check. Instead, I got a lecture, more or less, about wasting her time.

The attitude and tone of voice immediately threw me back into my own years of dating. Yes, ask men and we will tell you how we experience sexual selection, day in, day out. Nothing seems to make a woman angrier than being approached by a man she perceives as being inherently unworthy of her attention. Such men are wasting her time. Time is important, the reproductive clock is ticking and if you are a low status male, you are merely blocking access, obstructing a more worthy male who even now might be crossing the room to chat. Hence, you, my fellow male, you are not just invisible, you are getting in the way. It does not feel good to be told so, but that is reality. It might be worse still: the mere fact that a woman is seen speaking to a low status male is a potential threat to her because other people might see you with her, sending the unwanted signal that she is unworthy of attention from higher status males nearby. Hence, you need to disappear, perhaps even while she tells her friends loudly just how unworthy you are, in case that more desirable male is already watching and listening. If you think of it this way, the public shaming of low status men is actually a kind of evolutionary signal of high female value. Hypergamy, then, the logic continues, has the unhappy requirement that unworthy men must be quickly and cruelly dispersed, to open opportunity for the better man, higher in the hierarchy, no doubt hovering just out of sight, watching and waiting to start up a conversation.

The final anecdote. In order to do due diligence on female strategies and in order to escape routine work, I carefully read some online advice offered on how women can find the right man. There are, they say, plenty of fish in the sea. There is a huge library of online advice, woman-to-woman. The last story for today? It is unwise to ask a man how much he makes on the first date. (In any case, the columnist adds helpfully, if you are paying attention, you can usually guess it by looking at his car.)

So, there seem to be two unhappy realities at play. Men are sorted into hierarchies, and women prefer men nearer the top. Those serious principles are illustrated by my anecdotes and also explain why peacocks have such beautiful tails. And why finding a mate, keeping a mate, and raising children with that mate, continue to fill advice columns and sell books.

You may think that you can escape, like the Buddha, by leaving society and moving to the forest. Alas, even on a hiking trip in the mountains, some people demonstrate superior abilities in carrying packs and pitching tents, and it soon becomes evident who they are. I can also assure you that even when you live in the forest, as I do, life is still influenced by hierarchies, including the county council that maintains the roads, the corporation that delivers electricity, and even editors who decide which of my books and papers you will be able to read.

6
Killing Minds and Killing Fields:
Interference, Competition, and Aggression

> And so, if we are to be judged by the wishes in our unconscious,
> we are, like primitive man, simply a gang of murderers.
> (*Freud 1915, p. 765*)

THE ONGOING SLAUGHTER in the muddy trenches of the First World War was probably on Freud's mind when he penned this bleak observation in his 1915 essay, "Thoughts for the Times on War and Death." And after nearly a further full century of continual conflict, one can feel increasingly inclined to agree with him. What is the source of this unconscious urge to kill one another? Why does it elude diplomats and even the efforts of entire organizations like the United Nations? Why do new conflicts break out so routinely that many don't even make it to the front pages of newspapers?

There may be a simple answer. Recall that ape minds are constructed to crave the resources needed to build more apes with similar genes. From the pleasing appetite for fine wine to the covetousness for a neighbour's land, these and all the styles of feeling in between express a fundamental part of our ape nature. While much has been written about human aggression, far too often we fail to see that aggression itself is only a secondary state, a state driven by the hunger to acquire new resources (or mates). Although we have created a world with unparalleled wealth, these basic behaviours still haunt our unconscious, and still prevent us from getting satisfaction. We will look at some examples of just how much warfare is really driven by resources and the combination of conscious and unconscious desires to acquire them.

Ecologists have spent more than a century studying competition among organisms—all kinds of organisms, not just primates—for resources and now recognize two types of competition. In exploitation competition, neighbours interfere with one another solely by reducing the resource supply. For example, fishing communities may have negative impacts upon one another simply by reducing to overall stock of fish. Trees reduce the growth of neighbour trees by taking up water and nutrients they could use for their own growth. In interference competition, by contrast, neighbours deliberately and directly interfere with one another in order to control the resources. In the case of fishing, a community might damage neighbours' fishing gear or even burn their homes to drive them out of the region. There is less evidence for this in the plant kingdom, and it is likely that plants have to contend mostly with exploitation competition. But interference competition is an all-pervasive element of human history, and, in order to understand our own craving and aggression, it is imperative that we appreciate the power of this emotional state. Let us examine a few examples, ranging in time from ancient Rome to modern Europe.

Competition and warfare in ancient Rome

Julius Caesar understood well that competition between neighbours had two components: control of resources and direct interference with survival. In his campaigns in Gaul, Caesar describes the siege of Uxellodonum in 51 BCE, where he not only surrounds the town with earthworks, but builds a second ring of trenches with pitfalls and sharpened stakes facing outward to prevent any attempts at rescue by neighbouring Gauls. Learning from deserters that the town is well-supplied with grain, he decides to cut off the other essential resource: water. Roman soldiers deny access to the local river.

> The townsmen however still resisted obstinately and
> held out even when numbers had died of thirst,
> until in the end our mines tapped the rivulets which
> supplied the spring and diverted them. The sudden
> drying up of this supply, which had never failed them
> before, reduced the Gauls to such despair that they ....
> surrendered. (Handford 1951, p. 257)

Recognizing that military power might still interfere with the future control of resources, Caesar further records: "All who had borne arms had their hands cut off and were then let go, so that everyone might see what punishment was meted out to evildoers" (p. 257).

Strategic bombing, resources, and interference competition in humans

Strategic bombing campaigns of the mid-twentieth century provide another, more modern, illustration of the options available for interference competition with neighbours. The fundamental problem for modern military analysts is to decide which target will be most effective in reducing the enemy's power. Consider the Second World War, for which we have excellent historic records. One short account of the air war, *Bomber Offensive: The Devastation of Europe* tells how Bomber Command was instructed in 1941 "the sole primary aim of your bomber offensive, until further orders, should be the destruction of the German synthetic oil plants" (p. 32). It was not clear, however, that the accuracy of contemporary bombing technology could achieve this goal. To evaluate the option, aerial surveillance was carried out at two synthetic oil plants at Gelsenkirchen on Christmas Eve 1940. These two plants had been attacked by 196 bomber sorties, and 262 tons of high explosive, and an unrecorded number of incendiary bombs had been dropped. But neither target had apparently suffered major damage. Thus, whatever the best target in principle, there was still the practical problem of finding the target, dropping the bombs accurately upon it, and escaping with acceptable casualty rates. All of these latter tactical problems had to be borne in mind when considering the strategic objectives, and one by one these were solved by improvements in technology as the war progressed. Yet, strategic differences of opinion remained.

Even in the last years of the war, there were still three conflicting views as to how strategic bombing would most effectively interfere with the war capabilities of Germany. One view was that the destruction of synthetic oil factories would still be highly desirable because it provided the fuel for the entire enemy war machine, as well as for the industrial plants that produced more war machines. Moreover, this provided a fairly specific, if well-defended, target. Lieutenant-General Carl Spaatz, Commander of the US Strategic Air Forces favoured this target as did Sir Charles Portal, Chief of the British Air Staff. The second view was that transportation systems provided a highly visible target that would also hamper the movement of men and materials and as well as interfering with industrial production; it was so widespread that it could be attacked by many small bomber formations operating with relative independence. Air Marshall Sir Arthur Tedder favoured this approach. Finally, there was the view that, in the end, it was necessary to destroy the enemy's morale and infrastructure, and this could be done only through area bombing of entire cities. This would have the advantageous side-effects of also destroying industrial and transportation centers. Air Marshall Sir Arthur Harris took this view.

The relative expenditure of resources to these three interference strategies can be measured by the tonnage of bombs allocated to each. During the last three months of 1944, Bomber Command devoted 53 percent of its tonnage to area attacks on large German cities, 15 percent to railways and canals, 14 percent to oil plants, 13 percent to direct military targets in support of armies, and 5 percent to naval and other targets. This led to intense arguments between Portal and Harris.

Evaluation of these positions after the war ended has tended to favour the first option; shortages of aviation fuel and diesel had become critical by August, and the British and Americans were "gravely mistaken" not to have concentrated a greater proportion of their efforts upon it after the initial successes of May and June 1944. It is, says Frankland, virtually beyond doubt that the German economy would have been crippled if this target had been continually further attacked. This tends to reinforce opinions that the Second World War was primarily decided by the ability of each country to allocate industrial production to create weapons.

Historians also have evidence from the German perspective to evaluate. One example is the memoirs of Albert Speer, who was largely responsible for keeping the German war machine running (and was imprisoned after the Nuremberg Trials for twenty years, although the Russians voted for death). Speer says "As early as September 10, 1942, I had warned Hitler that the tank production of Friedrichshafen and the ball-bearing facilities in Schweinfurt were crucial to our whole effort ... the war could largely have been decided in 1943 if instead of vast but pointless area bombing the planes had concentrated on the centers of armaments production." (Speer 1970, p. 365) He also adds that the German Luftwaffe should have adopted a similar strategy, and even set up a committee to select these targets.

> I shall never forget the date May 12 [1944].... On that day the technological war was decided. Until then we had managed to produce approximately as many weapons as the armed forces needed, in spite of their considerable losses. But with the attack of nine hundred and thirty-five daylight bombers of the American Eighth Air Force upon several fuel plants in central and eastern Germany, a new era in the air war began.

> It meant the end of German armaments production. The next day ... we groped our way through a tangle of broken and twisted pipe systems ... even optimistic forecasts could not envisage production being resumed for weeks (p. 445).

The military examples emphasize two principles that may have wider application. First, interference competition will be most effective when it targets a critical limiting resource for a neighbour or when it damages its ability to transport that resource. Second, whether in organisms or human societies, energy is often the most important resource, if only because energy reserves measure both the ability of a neighbour to forage for new resources and its potential to create structures for interference in the reverse direction.

An apparent digression: competition among plants

Plants are a useful contrast to humans, because, so far as we know, they are unconscious entities. Therefore, if we take a break from human destruction and consider the perspective of plants, we can treat competition as an entirely passive process driven by the evaporation of water from the soil into the atmosphere—a process that occurred long before there was life at all. Plants occupy the precise interface where this evaporation occurs, increasing its efficiency, so much so that biologists now describe this step in the hydrological cycle as evapotranspiration. Water is pulled from the stomata in leaves by evaporation, and this draws water upward through the tissues of the plant; as the upper tissues of the plant dry out, osmotic differentials draw water out of the roots. As the roots dry out, water from the soil is drawn in. Plants therefore occupy the site of phase transition from liquid water to water vapour. As the sun withdraws water vapour, mineral nutrients are pulled out of the soil, conducted through the xylem of the plant, and used in the manufacture of carbohydrates and proteins. Smaller amounts of water are actually consumed in the process of photosynthesis, releasing the waste product oxygen back into the atmosphere. This is a difficult perspective to retain, in part because biologists have been trained to think of populations and individuals as the natural units of the living world, and humans in general tend to view the world as a composite of individuals. But if we maintain this anonymous perspective of plants as largely passive conductors of water to the atmosphere, we gain a fresh perspective upon the process of competition. From this perspective, some plants, by virtue of their size or location or internal resistance, passively transport more water to the atmosphere. Others, because they are exposed to less sunlight, or occur in drier soils, or are found in depletion zones created by neighbours, or because they inherited higher resistance to water flow, transport less water to the atmosphere. The physical processes of diffusion ensure that, in general, the larger individuals conduct a greater flow of water, and so they withdraw more nitrogen and phosphorous from the soil, and consequently, during photosynthesis, construct more proteins and carbohydrates. These in turn increase the evaporative area of the plant, which increases the rate of transpiration, which increases the rate of passive evapotranspiration.

A little more biology must be spliced in here to ensure that the plants with the larger rates of evapotranspiration construct roots to provide the necessary resources; one can imagine even some simple feedback loop whereby those roots that conduct the greatest flow of water and nutrients upward are provided with the most return flow from the branches. Indeed, in many plants individual roots can be associated with individual branch systems in just this way, so a tree may be regarded a set of weakly interconnected shoots—a community of branches—each somewhat isolated from the other and each competing within the tree for access to light and water. Those shoots that lie, for example, low on the tree where they are shaded, die, and these weakened branches fall to the ground, whereas other shoots that are exposed to conditions where they can grow rapidly, attract more water from within the plant, thereby growing larger. Thus, trees passively respond to light and moisture gradients created by the sun and surface of the planet, and a natural and largely passive response results in some shoots and meristems multiplying at the expense of others. A plant can be viewed as a mechanical evaporative structure that enhances the flow of water from the soil to the atmosphere. Competition is a word we apply when we see that some of these structures, which we call individuals, and which we assume to be genetically distinct, increase by diverting resources from their neighbours.

The looting of Europe in the 1940s

Now back to humans. Humans seem to have a strong, innate desire to take items that belong to other humans. There is a tendency to think that modern humans and modern war are different, and much of the popular writing about the Second World War concerns military events. Let us briefly consider the extent to which it was driven by simple pillage. In *The Rape of Europa*, Lynn Nicholas describes how the Nazis systematically pillaged conquered territories, with military victories swiftly followed by teams with lists of which art was to be stolen and taken back to Germany. Similarly, the destruction of European Jews in the Holocaust also had an element of pillage: the houses and furniture of the deported Jews were available for occupancy and disbursement to Germans. If you have always wanted your neighbour's house, or their art, or their furniture—denounce them to authorities, and it will be yours.

Here is a personal account from later in that same war, and it describes items collected not by high-ranking Nazis, but by ordinary soldiers. Our witness is Alan Moorehead, a well-known foreign correspondent for the *Daily Express* in London. *Eclipse* is about his travels with the allied armies, from August 1943 to May 1945. He was there to witness the invasion of Italy, the landings in Normandy, and finally, the fall of Germany. Moorehead describes how trains were packed with loot. "The Nijmegen railway yards contained a train full of presents the German soldiers were sending home to their families."

> I opened packets containing such things as a gross of curtain rings; sets of spanners and electrical tools; barbers' clippers by the hundred; hundreds of pocket-knives and bottles of varnish. These were addressed to young unmarried girls in Germany. Heaven knows what Fräulein this or that was supposed to do with these cases of false teeth, the coils of fencing-wire and the little boxes each containing a thousand gadgets for fixing pictures on the wall. It was the sentiment I suppose that counted. One lucky girl would have been the happy recipient of ten chisels. Another was being offered forty-six hammers and a case of unbleached cardboard. (p. 213)

As described in *The Rape of Europa*, pillaging had been ongoing with orders from the highest levels including Hitler and Göring, ever since the Nazis conquered their first territories. It ranged from organized teams heading for the art galleries to some of the individual acquisitions of soldiers described by Alan Moorehead. The train he describes was just one of hundreds that had passed that way.

Bertolt Brecht actually wrote a memorable song about this very process, "The Ballad of the Soldier's Wife." It starts with a German soldier sending his wife a pair of high heeled shoes from Prague and then a collar of fur from Oslo, both cities that were early Nazi conquests. The song continues with a hat from Amsterdam, lace from Brussels, a silken gown from Paris, and an embroidered shirt from Bucharest. The music was composed by Kurt Weill, who, like Brecht, was forced by the Nazis to flee Germany in the 1930s. A haunting version is sung by Marianne Faithful on an album called simply *The Music of Kurt Weill*. The song also raises another uncomfortable issue: the degree to which women are complicit in pillage, since they enjoy some of the benefits. According to the song, the wife is happy with all her foreign gifts except the last one: a widow's veil from Russia.

We should be clear that the example of the Nazis in the Second World War is just that—one example, selected because it is relatively modern. Another well-known example comes from the previous century, in 1812, as Napoleon and his troops pillaged Russia and particularly Moscow. The retreat from Moscow left roads littered with valuables that had been discarded by retreating French troops. Napoleon himself had an entire wagon train filled with valuables, and when these wagons were eventually abandoned, their contents were in turn picked over by retreating soldiers. According to *Moscow 1812: Napoleon's Fatal March*, "The ground was littered with silver coins and other booty being cast aside as the men filled their pockets and knapsacks with gold, jewelled icons, and other pieces of Napoleon's Moscow booty." (p. 514) Meanwhile, survivors describe how the freezing and retreating French soldiers were reduced to wearing every scrap of clothing they could find—including the dresses and furs they had collected for their wives.

Kleptocracy and the de-industrialisation of Africa

Since many of the examples of war and pillage so far come from Europe, let us consider two other examples. Human history abounds in these behaviours, whether in the indigenous cultures of North and South America, Africa, or Asia. To re-emphasize that it is not just twentieth century Europe that was somehow different, we will briefly consider examples from contemporary Africa and then ancient Hindustan.

Consider the region of Africa that includes the Congo River watershed. It was once called the Republic of Zaire and now it is the Democratic Republic of the Congo. There are many valuable natural resources in this region of Africa, including diamond, gold, copper, and cobalt mines. This region is also globally important for conservation: there are the immense wetlands of the Congo River floodplain, extensive tropical forests and wild populations of our closest living relatives, chimpanzees, bonobos and gorillas. All are at risk from pillaging. For some decades, the leader and dictator of Zaire, was a former sergeant major in the Belgian Congolese army, Mobutu Sese Seko, who distinguished himself with a form of rule that has come to be called a kleptocracy.

In a kleptocracy, a leader systematically loots his own country. Resources flowed from the mines and forests through Mobutu and into foreign bank accounts, allegedly making him one of the world's richest men. Meanwhile, the infrastructure of Zaire decayed; forests even reclaimed major roads. If we consider how resources flow into plants, as described earlier in this chapter, we can view this period of Congolese history as just another illustration of the passive flow of resources from a source (forests and mines) to a sink (bank accounts and villas), a financial flow that is in many ways as natural and passive as the flow of water from the soil to the atmosphere. Adjoining branches, that is political allies in the *Mouvement Populaire de la Révolution*, benefited as the resources flowed past them toward Mobutu. Rival political groups who would direct the resources to other branches were ruthlessly suppressed.

A guerrilla army that arose in the east of the country near the Rwanda border in 1996 slowly cut into Mobutu's resource base; wealth that once flowed into Mobutu's depletion zone then flowed eastward into the rebel army's depletion zone. Several months passed and Laurent Kabila came to power. Soon he was as deeply reviled as Mobutu and by some accounts equally given to kleptocracy. Now the country is ruled by his son Joseph who has shown an equal reluctance to stand for election, in spite of being called a President. Elections were scheduled for 2016 but did not occur. After the elections were finally held in 2018, Kabila still managed to hold on to his power in spite of being defeated by a rival.

Overall, one cannot help but think of plants, in which one branch simply replaces another on the tree trunk, and the winner is distinguished by having the resources trickle through its own tissues rather than those of a neighbour.

Hierarchies and plunder by Muslim armies in India

And now India, more than one thousand years in the past, then known as Hindustan. Well after the Buddha, but well before Darwin, Islamic armies invaded India in three waves under three leaders, Muhammad bin Qasim (ca. 712), Mahmud of Ghazni (ca. 1000) and Muhammad of Ghauri (ca. 1175). The Indian historian Kishori Lai wrote at length on this topic. He observed that all three waves of invasion involved similar behaviour. All males who survived the conflict were killed unless they converted to Islam. Women and children were taken as slaves, with the most desirable women being returned for rulers higher in the chain of command. And there was widespread pillage. Indeed, Lai's material is difficult to read because of the sheer repetition: it is easy to lose track of which unfortunate city is under siege.

Regarding slaughter, here is one account from the attack on Thanesar, recorded by a contemporary Muslim writer " …The blood of the infidels flowed so copiously that the stream was discoloured and people were unable to drink it … Praise be to Allah … for the honor he bestows upon Islam and Musalmans."

These were not isolated incidents. There was a well-planned philosophy of conquest. Lai quotes Tamerlane's own words on his plans for India.

> My principal object in coming to Hindustan … has been to accomplish two things. The first was to war with the infidels, the enemies of the Mohammedan religion; and by this religious warfare to acquire some claim to reward in the life to come…. The other … was that the army of Islam might gain something by plundering the wealth and valuables of the infidels; plunder in war is as lawful as their mothers' milk to Musalmans who war for their faith."

The wealth that flowed back to Muslim leaders included gold, silver, jewellery, and precious stones. Other valuable goods included indigo, embroidered silk, cotton, and Indian steel. "More and more wealth was drained out of the Punjab and other parts of India. Besides the treasure collected by Mahmud, his soldiers also looted independently." There is also a substantial amount written about the institution of slavery and the flow of slaves outward from Hindustan into Muslim nations.

These historical events have particular significance to Buddhists. For many years teachers have explained to western students how Buddhism almost disappeared in its source country, the land of the Buddha himself. Our knowledge of the Buddha's teachings therefore mostly comes from other parts of the world, particularly the mountainous country of Tibet and the island now known as Sri Lanka. What happened to Buddhism in the land of the Buddha? And what happened to famous Buddhist institutions that are mentioned in texts and liturgy—monastic communities such as Nalanda and Vikramshila? According to *Land of Seven Rivers*, a history of India, both of these latter institutions were attacked and plundered by Muslim armies led by Muhammad bin Bakhtiyar Khalji in 1202. The Buddhist monks were slaughtered. Historians tell us that he returned with great amounts of plunder and was amply rewarded by the caliph Qutbuddin Aibak, the founder of the Mamluk dynasty.

Dominant males and floaters: dangerous young men

Now let us change our focus to consider another perspective on pillage. Within human societies, as within tribes of other primates, there is competition for resources both within a group and between groups. Young males pose a particular challenge, particularly if there is no obvious means to advance in society. Consider the situation within a human society. There can be competition within a tribe (for access to land and inherited wealth and mates) and as well as between tribes (for access to new land and plunder and mates). In the middle ages, bands of young men without land were a constant social problem and source of conflict. *A Distant Mirror*, by Barbara Tuchman, provides many examples from the 13th century. Since these young men have no territory, sociologists like James Boone have termed them 'floaters'. This is an interesting use of words, since birds are also strongly territorial, and ornithologists call young males that lack territories floaters. These subordinate floaters provide a source of conflict. One medieval solution was to arm them and send them somewhere else. "Vagabondage was not usually solitary, and often bands of 'youths' lived and travelled together: a newly dubbed 'youth' might arm and take with him the sons of his father's vassals who were his own age." In those times, the word youth might be applied equally to older males who were unlikely to inherit land. Boone therefore suggests that some wars in the middle ages were started by older males who were trying to absorb the envy, energy and threat posed by younger, landless males.

We could add a Darwinian perspective to the observations of those of the sociologists. In his explanation of warfare as a means of minimizing the risks of within-group competition within a social hierarchy, Boone seems to overlook the fact that rape, pillage and enslavement may have important evolutionary benefits to the existing hierarchy, in the dispersal of selfish genes to new habitats. These dispersing males may even become the source of new lands, new resources, and more females to perpetuate the genes in the landed hierarchy. If we re-name floaters to 'dispersers', it might change the way we look at such activity. In any case, it is important for us to consider the possible biological basis of historical events.

A more recent example of young men released from normal social constraints comes from southern Europe, and what used to be called Yugoslavia. Michael Ignatieff has described events there. We can see the ancient dominance hierarchy of Chapter 5, and the concept of young male floaters, erupting into modern events.

> The history of our civilization is the history of the confiscation of the means of violence by the state. But it is an achievement that an irreducible core of young males has always resented.... As everyone can see on his television screen, most nationalist violence is perpetrated by a small minority of males between the ages of eighteen and twent-five, young males intoxicated by the power of guns on their hips.... I met lots of young men who loved the ruins, loved the destruction, loved the power that came from the barrel of their guns. (Ignatieff 1993, p. 246-247)

These bands of young males, says Ignatieff, are led by warlords, who appear "wherever nation-states disintegrate: in Lebanon, Somalia, northern India, Armenia, Georgia, Ossetia, Cambodia, the former Yugoslavia." (p. 40)

"Their vehicle of choice is a four-wheel-drive Cherokee Chief, with a policeman's blue light on the roof to flash when speeding through a checkpoint." (p. 40) They dress in leather jackets, floral ties, and pressed corduroy trousers. Arkan, for example, controls an eight-hundred strong paramilitary unit called the Tigers, "who raped and tortured their way through eastern Slavonia in the Croatian war of 1991." (p. 41)

Thankfully there is now some measure of peace in Yugoslavia. But if you look carefully at photographs of the migrants now arriving in Europe, you can see that a great majority of them are young men. Where are the families? It is difficult to get accurate information. Indeed, if you know something about evolutionary history, it is worrying to see that even when these migrants are nearly all young men, reporters seem to deliberately seek out young women with children to photograph, and it is these highly selective photographs that appear in the news. Since the photographs do not seem to accurately reflect the actual sex and age of migrants, it would appear that even the reporters are aware that bands of young males represent a threat to social order.

Becoming aware of how violence emerges from the unconscious

Let us now take a break from history. The reason for dwelling on war and pillage was to make the point that the capacity for this behaviour seems to be an inherent part of our primate nature. Hence, it is a significant challenge to build a world in which this potent force can be accommodated. But we must be realistic. Those who pretend that human aggression does not exist will end up like the ineffectual European diplomatic teams who were sent to Bosnia (a part of the former Yugoslavia) and who extracted useless promises from the Serbian warlords. Repeatedly. Or possibly like the unfortunate Neville Chamberlin, the British Prime Minister who is now remembered mostly for his pronouncement, after a meeting with Hitler in 1938, that England and Germany would never go to war again. We know how that turned out: war broke out the very next year. The most dangerous forces in our psyche may be those that we refuse to acknowledge.

J.R. Saul has addressed the need for this kind of self-knowledge. He offers a diagnosis remarkably reminiscent of the Four Noble Truths. Who is more contemptible, he begins, than he who scorns knowledge of himself?

> We have difficulty in perceiving our own weaknesses.... If we are unable to identify reality and therefore unable to act upon what we see, then we are not simply childish but have reduced ourselves to figures of fun—ridiculous victims of our own unconscious.... If we cannot see ourselves, then we cannot act as humans. (Saul 1995, p. 23)

So what might the Buddha have to say? He might remind us of the importance of self-examination. The Buddha's meditation instructions give us a means to look more carefully at the thoughts and emotions that arise in our minds and cloud our judgment. In order to do so, he tells us, we need to cease judging our thoughts as either right and wrong, or good and bad, but rather we can allow them to pass by in all their display. As they pass by, we can examine them without bias, and even begin to make friends with them. They are only thoughts. In this manner, the power of illusion loses some of its evolutionary grip upon our motivation. Once thoughts and feelings emerge into the conscious, we can see their hands grasping at the controls of ape ego. To paraphrase *The Dhammapada*, "Good and bad or happy and sad, all thoughts will vanish, on their own, like the imprint of a bird in the sky."

Self-examination is likely a good thing for humans. It is also possible to misunderstand the process. Indeed, there appears to be misunderstanding by some contemporary Buddhists. The basic instructions of meditation are to allow thoughts and feelings to arise without judgment. Usually this is done while seated, but there are also various styles of walking meditation. We set aside time to directly experience our own inner monologue. It is by no means always a cheerful process. The inner monologue sometimes seems like a collection of film fragments, some of which are quite irritating and even horrifying. Clearly, the unconscious mind is a busy place.

And here is where some Buddhists seem to miss the point. I have heard certain Buddhist teachers saying that we should let go of right and wrong, and certainly some students hear that there is no right and wrong. The assertion that there is no right and wrong is a strange and dangerous (and false) interpretation of Buddhism. Letting go of the concept of right and wrong is, however, an essential tool when one is sitting on a meditation cushion. It is just not advice for running a modern society. The Buddha's meditation instructions are not meant to be political philosophy. Possibly some of the confusion arises because students do not always accurately hear what they are told. (Anyone who has given excellent lectures and then marked exams, knows how often students miss the point.) Also, one's ability to understand instructions grows with experience, just as with a musical instrument. So one frequently requires many periods of instruction and dharma talks, mixed with practice. It is evident that in applying Buddhist teachings to modern times, there is some need for discernment, as well as some sort of familiarity with history and human nature.

It seems that modern people have a challenge. We must somehow chart a middle way between the two extremes of despair and wishful thinking about our inner nature and our evolutionary baggage. Charting this course will require what the Buddha called skillful means. In this context, skillful means can be understood to be a fuller understanding of the constructive and destructive forces within the human psyche. Even the Buddha had to warn his students to 'abstain from taking what is not given'. There seems to be an important and obvious lesson. If a political or religious leader says they plan to kill you and take your belongings, there are strong reasons to assume that they mean what they say. They are invoking what appears to be one the most powerful and ancient evolutionary forces that lies buried within our human nature. Given our evolutionary drives and the historical precedents, it seems that we will have to remain continually aware of those around us who would do us harm. We come from ancestors who perfected the art of taking what was not theirs and will have to live, perhaps uneasily, knowing that this too is part of human nature.

7
Insatiable Consumption:
When Big Brains Meet Big Animals

> We are weary of witnessing the greed, selfishness and cruelty
> of 'civilized' man toward the wild creatures of the earth. We are sick
> of tales of slaughter and pictures of carnage. It is time for a sweeping
> Reformation; and that is precisely what we now demand.
> (*Hornaday 1913, p. x*)

THERE ARE IMPORTANT TOPICS on which the Buddha is strangely silent. It is time to look at one: our relationship with wild nature. The Buddha and other important Buddhist teachers like Shantideva give wild nature scant attention, except to advise people to meditate in the forest. Darwin, on the other hand, paid close attention to wild animals and wild landscapes. Whether it was tortoises or barnacles, he appreciated their diversity, took time to observe them closely, and wrote about them for a wide audience. *The Voyage of the Beagle* is a classic example. The opening manifesto of this chapter was penned by a biologist who was born just five years before Darwin published *The Origin of Species*. At the time William Hornaday was writing, in the early 1900s, wild birds were being slaughtered just so women could wear hats adorned with colourful feathers. Other wild birds like robins and passenger pigeons were being shot in their millions as a source of meat for New York restaurants. It was mostly biologists, not Buddhists, who took action to protect those wild species, and the wild places in which they lived. So, as we walk the trail with Darwin and the Buddha today, let us have a discussion about this difference in attitudes. We saw in Chapter 1 that all humans experience dukkha (suffering, dissatisfaction), and this is a root cause for the destruction of wild nature. We will look at the impacts of humans on wild nature, focusing particularly on large wild animals. We can be sure that today, on

Giant ground sloths once ranged through North American landscapes, but are now extinct. They were one of the largest land animals ever to live, weighing more than modern elephants. North America once supported an entire fauna of similar magnificent animals—mastodons and woolly mammoths, saber-toothed cats, camels, wild horses, giant tortoises, and giant beavers. The most probable explanation for their demise is extermination by ancient human hunters. Similar destruction of large wild animals by humans, long before the arrival of European explorers, is documented in Australia and New Zealand.

the trail, we will see neither giant ground sloths nor passenger pigeons. Their absence will guide the conversation.

Humans and wild nature

Although we think of ourselves as modern people, driving cars and flying in spacecraft, our brains and our consciousness are firmly rooted in the past. The plains of Africa were our own Garden of Eden. We hunted animals, and we gathered plants, and our minds remain a product of this experience. For example, we have two eyes (to help judge the distance to prey) and colour vision (to help detect ripe fruit.) A key part of this evolutionary interbeing is our relationships with other large animals. How humans relate to large animals tells us something about how humans relate to wild nature. It turns out that large animals are repeatedly over-hunted, often to the point of extinction. We are the principal cause. The reason lies in the first chapter. Humans crave resources, and meat is one such resource. Humans also have a strong drive to produce more copies of themselves. Hence, for millennia, we have been busy turning wild animals and wild nature into more people. Here is just one recent example. The 2018 *Living Planet Report* by the World Wildlife Fund tells us that in the past forty years, there has been a 60 percent decline in wildlife populations. (A particular case: in 2018, while I was working on this manuscript, the last male Northern White Rhino died.) Meanwhile, in just the last twelve years, human populations have climbed by another billion.

In the past, stone-tipped spears and wooden digging sticks limited the damage done. Rifles and roads make it far easier to damage wild nature today. When we lived in small tribes, there was likely little natural selection for self-restraint. If you over-killed locally, and there was no wild game left, you simply moved on. Now that we can no longer move on, we need to look more closely at our destructive desires. It seems that there is no natural constraint on our capacity to kill. We will look at some examples, then we will look at what the Buddha said, or did not say, and then we will consider where that leaves modern humans.

I think it is best to give you one conclusion in advance. The human craving for meat in particular, but natural resources in general, has always had dangerous consequences for other species. When the human population was small and scattered, the effects were less obvious. As human populations grew, the impacts of humans grew as well. This pattern crosses cultures and races: the evidence suggests that aboriginal people can be every bit as destructive as contemporary people.

The disappearance of North America's megafauna

North America once supported a fauna of magnificent animals that rivalled those of Africa. There were mastodons and woolly mammoths, saber-toothed cats, camels, wild horses, and giant ground sloths. The ground sloths were one of the largest mammals ever to live, weighing more than modern elephants. In addition, there were beavers the size of small bears, vultures with twelve-foot wing spans, giant armadillos, and even giant tortoises.

Today they are vanished. Only a few large animal species such as bison, moose, and muskoxen remain in North America. How did all the others disappear? At first we might blame the four ice ages that have swept down over the continent during the last million years but there is now good reason to exclude this explanation. Fossil remains tell us that many of these vanished species survived the great ice ages, only to disappear within the last 10,000 to 20,000 years, after the last ice age had ended! How could a wide array of animal species survive four ice ages and then vanish so suddenly?

This extinction of North America's largest animals coincided with the arrival from Asia of humans in North America. Some biologists propose that our megafauna may have vanished under an onslaught of hunting by the Clovis culture which was widespread in North America at that time. Needless to say, this theory was at first controversial. Others have argued that aboriginal peoples of North America did not have the technological capacity or population density to cause these extinctions. Let us consider the options. More recent estimates of the human population in North America suggest that it had perhaps reached some 60 million at the time Christopher Columbus arrived. Of course, this population size was achieved more than ten thousand years after the Clovis culture who hunted the megafauna but the point is that there were a lot of people in North America in the past. Recent archeological excavations also show that stone tools are closely intermingled with the bones of large mammals at multiple sites in North and South America. Although there are still some scientists who think natural climate changes after the last ice age should still be considered as a possible cause, I consider this to be the least plausible cause of extinction because these animals tolerated the changes in climate and vegetation of many previous ice ages. To save this hypothesis, you would have to argue that there was something entirely different about the last ice retreat. What was certainly different was this: humans had arrived from eastern Asia and were now hunting large animals.

What seems to have been overlooked in this debate is one important point: it does not matter whether or not the last giant beaver or giant ground sloth was actually speared by an aboriginal hunter. The fact remains that an indigenous human culture existed at that time. North America was occupied by humans. Let us apply the same test to them that we apply to my own modern culture. Did anyone notice the fauna disappearing? Did anyone care? Did anyone do anything to try to stop it? More significantly, since such questions may never be answered, let us ask one question only—did they succeed in preventing the extinctions? The list of vanished species speaks for itself. Major failure. The wave of extinctions at that time exceeded anything North America has since experienced. Hornaday is writing about a second wave of destruction, one that was mostly aimed at smaller animals. It is probably no exaggeration to say that aboriginals presided over the greatest failure of natural resource conservation in the human history of North America.

There is one remaining possible counterargument. The final disappearance of the megafauna has now been pinned down quite tightly to 12,900 years ago. Findings of stone tools and megafaunal skeletons end rather abruptly at this quite sharply-delineated time. The timing is coincident with a period of cooler weather worldwide, known as The Younger Dryas (named after a flowering plant *Dryas octopetala* that became more common in Europe during this period). Although the Earth was emerging from the last ice age, this was a brief return to what some have described as near-glacial conditions. There is some evidence that an asteroid or comet may have struck the Earth at this time, triggering an abrupt change in climate and possibly terminating both the megafauna and the culture of their human predators. It will take further effort and many more excavations to collect evidence for and against this hypothesis. Perhaps it was a minor version of the great asteroid collision that ended the era of dinosaurs some 60 million years earlier. But even the proponents of the extraterrestrial impact hypothesis say only that this event may have 'contributed to' the megafaunal extinctions, as the intermingled stone tools and bones show that the slaughter was already in progress. In such circumstances, a meteor or comet may simply have provided the final blow.

Global defaunation

Perhaps North America was an unfortunate exception. Paleoecologists, however, have now gathered evidence showing that rather than being unusual, this pattern was typical of the effects of aboriginal peoples upon new lands. There are many examples. The Maori, for example, reached New Zealand from Polynesia about 1000 years ago. They found great flightless ostrich-like birds called moas in great numbers. By the time Europeans arrived, moas survived only as faint memories and bones, although some are of the opinion that they did not finally die out before Europeans could at least, in theory, have seen a few. Similar waves of extinction swept over Australia. It appears then that most of the large animals on these continents had been eradicated before Europeans arrived.

This process has not ended. The peoples of Africa and Southeast Asia are now destroying the last animal populations on an industrial scale in what is known as the bushmeat industry. Although humans have hunted animals in tropical forests for millennia, the intensity and area of bushmeat hunting is now unprecedented. It is a perfect storm driven by growing human populations, cheap guns, and roads. According to a recent study led by biologist Rhett Harrison, "Over the past 2–3 decades, the availability of firearms and affordable ammunition, on the one hand, and improved access to forests and markets, on the other, has resulted in widespread declines in wildlife populations throughout the tropics" (p. 687). There is also the added factor of human population growth in tropical areas. Consider just one example, involving just one road in Borneo, a part of the world that has animal species found nowhere else. Over 15 years a team tracked the situation in one forest area. The onset of heavy hunting can be traced to a single access road that was upgraded in 1987. Within a decade (by 1998), hunters had exterminated all the animals larger than one kilogram, including hornbills, gibbons, and flying foxes.

Bushmeat hunting is an enormous threat to tropical forests, since over-hunting affects far larger areas of forest than deforestation and logging combined. Every access road can open a forest to defaunation. Roads are notorious for their negative impacts upon wild areas, and may be regarded as one of the most dangerous human activities in wild areas. Hence, maintaining roadless areas is an important global conservation priority. Some biologists are calling these defaunated green forests 'green deserts': the trees are still there, but the animals are gone. Even our closer living relatives—gorillas, apes, and bonobos—are being killed for their meat.

Shockingly, bushmeat is not only being consumed locally. It is being smuggled to immigrant communities in developed countries as a luxury, at high prices. A single seventeen-day investigation at the Paris Roissy-Charles de Gaulle airport found 188 kg of bushmeat being smuggled in personal baggage, which would mean 270 tonnes per year passing thorough just this one airport! It is disturbing to accept that even when people migrate out of Africa, their demands for wild meat continue to drive the process of wildlife extinction on that continent.

The collapse of aboriginal populations from environmental abuses

Archaeologists and ecologists have become increasingly interested in reconstructing the environmental history of entire civilizations. An issue of *Quaternary Science Reviews* reports on the history of the Mayan civilization in Central America. The data show that over a period of some 4000 years, Mayan populations increased rapidly, forests were cleared, soil eroded into lakes, phosphorous levels in the lakes rose dramatically, and then the civilization collapsed. Ecologists like me are particularly interested in such studies because disturbingly similar trends are found in our technological societies. Deforestation, soil erosion, and eutrophication are all problems in North America today. There is no evidence that the Mayans were any better than we are at protecting either forests or the water quality of lakes. Indeed, it seems they reached an ecological end point.

A similar story has now been reconstructed from Easter Island, the site of those famous stone heads staring out to sea. The story is one of population growth, deforestation, soil erosion, and ecological collapse. The Easter Island story is perhaps sadder because one can actually document how even essential life support species were driven to extinction; as remarkable as it may seem, a sea-going people destroyed their forest to the extent that they were unable to find trees large enough to make boats. Even the very species of tree once favoured for boat-making became extinct. This civilization then fell into a period of chaos and war. Huge effort was, however, invested in constructing and transporting stone heads. The aboriginal populations had collapsed even before Europeans arrived. This solves the mystery of who made the stone heads, and it seems that if they could speak, they would have much to say about how humans have difficulty managing forests and setting priorities for action.

The cod fishery and environmental attitudes in outports

The evidence is fairly clear that stupidity is not restricted to any single race. Consider my own experience with an indigenous white culture that has existed in the outports of Newfoundland for probably at least 500 years. Here was a civilization that had always depended upon the ocean as its economic lifeblood. Like so many others, I confess that I expected noble savages who loved and respected nature and the ocean. Consider these vignettes from a visit I made in the mid-1970s:

> After adding oil to the engine of our motorboat, our host tossed the oil can overboard. It sank to the bottom of the harbour ... which turned out to be covered in such oil cans. Closer to the shore the oil cans were obscured by soft drink cans.

> Our host took us to sea. He brought along several green plastic garbage bags which were tossed overboard as soon as we were out of the harbour. The shoreline, it turned out, was littered for miles with washed up debris from the outport.

> The peat bogs near our cabin were torn up for miles by all-terrain vehicles. The woods and shorelines were littered with tin cans and bottles.

> An acquaintance who worked on a fishing boat told me how fisherman would haul in cod, cut out their tongues to eat, and simply throw the living fish back into the ocean to die, tongueless.

These scenes of blatant environmental disrespect come from a people whose livelihood is the ocean. If I burned my books, smashed my computer terminal and threw garbage into my filing cabinets, people would call me crazy. As a consequence of such attitudes, the cod fishery has collapsed. The ocean it turns out, is not limitless in its ability to absorb abuse and provide endless fish. Of course, the story involves many more players than just Newfoundland, but today we are discussing the topic of attitudes and awareness.

Homage to Guy Bradley

Since the topic of the decline in populations of wild animals is an unhappy one, let me end with a positive example, an example of one person who really did make a difference. Because of him, you can still see egrets in the wetlands of southeastern North America. He is not a name well-known in history. He was Guy Bradley.

Today the stately great egret is a fairly common sight in the wetlands of the United States. Egrets can even be seen foraging for frogs and insects in ditches along the busy interstates. Although we now take these common birds for granted, they barely survived into the twentieth century. In the late 1800s, plume hunters were destroying the birds. The feathers were sold in New York and London to adorn women's hats. Look at paintings from the late 1800s and notice how many women are indeed wearing plumes and birds on their hats. So, you can see from history that the destruction of wildlife is not a gender-specific problem.

Shipments of bird feathers went to auctions in New York or London, just as furs do today. Smaller birds, such as warblers, finches, and hummingbirds, were sometimes skinned and used whole. William Hornaday, already mentioned, describes a shipment sold in London, England that included

> Half a million birds or parts of birds ... Osprey plumes, 11,352 ounces; vulture plumes, 186 pounds; peacock feathers, 215,051 bundles; Indian parrots, 228,289 bundles; ... tanagers and sundry birds, 38,198 bundles; humming birds, 116,490 bundles; jays and kingfishers, 48,759 bundles; ... owls and hawks, 7,163 bundles. Similar sales frequently take place.

The best feathers sold for more than $45 per ounce in 1913 dollars. In 1911, 21,528 ounces of aigrette plumes were sold by the feather trade; at six egrets to one ounce of feathers, that represents 129,168 egrets killed for just nine months, in just one city. With such demand, "plume hunters" methodically went to one colony after another, shooting the breeding birds while their feathers were at their best. The young birds were left to starve. In Florida, "All known rookeries accessible to plume hunters had been totally destroyed."

Several small remnant colonies were discovered. One colony was in Louisiana, on Avery Island. Another colony was found in Florida, on Oyster Keys. The newly-formed Audubon Society decided to act to protect this Florida population. They hired Guy M. Bradley to be a game warden. He was deputized by the Monroe County Sheriff's Office. On 8 July 1905, Bradley approached Walter Smith who, with his son and a friend, were killing egrets at Oyster Keys rookery. No one knows exactly what was said, but Bradley was shot and killed. Two shots were heard. Bradley's pistol showed no evidence of having been fired. The grand jury failed to indict Smith, and he was released.

Bradley was buried on a shell ridge at Cape Sable overlooking Florida Bay. His grave was later washed away in a storm. But his legacy is remarkable: today egrets wade by their thousands in our wetlands. A few years ago, one even visited a wetland in my own forest all the way up in Canada. And, I have done my own part to continue his legacy by buying these wetlands (a forty-year process) and ensuring that they are fully protected by our local land trust. So, single individuals can make a difference.

What Shantideva does not say

Shortly after laying out the Eightfold Path, the Buddha also began to set certain rules for human conduct. According to *The Historical Buddha*, his very first rule was to refrain from taking life. Some Buddhists indeed are vegetarian, although Gautama himself seems to have been a pragmatist. Monks were allowed to eat meat if it was offered during an alms round, so long as the animal was not deliberately killed for that purpose. Monks were also not allowed to damage plants. So, from one perspective, the Buddha did say that individual animals and plants should be treated with respect. Given the times, with hunting a popular sport, this was a positive act on his part.

However, there is an omission: the topic of wild landscapes and wild animals. What seems to be lacking, at least in the many modern talks I have heard and the books I have read, is a bigger view: a statement on the importance of respecting populations of wild animals and the wild landscapes in which they live. Since the Buddha's time, vast areas of Indian forest have been cleared, putting many wild species at risk. The long list of endangered species of India includes Bengal tigers, Asian elephants and the greater one-horned rhinoceros. Forest clearance continues, being driven in part by expanding human populations. The Ganges River basin, home territory of the Buddha himself, has lost some eighty-five percent of its original forests. Even the Himalayas that feed water into the Ganges are now threatened by deforestation. If the Buddha had made a statement about non-harming of wild places in general, and forests in particular, it might have made a difference.

Let us take a closer look at this issue using a well-loved book, *The Way of the Bodhisattva*, a guide to the spiritual life. Its author, Shantideva, was a Buddhist monk in India who was born in 685, long after the Buddha had died. Shantideva's book has been a bedside companion in my own life, and it is highly regarded by many Buddhists. Shantideva talks about the importance of retreating from the human condition and entering the forest. Naturally, I wanted to know more on what he had to say about doing so, and so I returned to the book to enquire about forests. There is not much, it turns out. Most of the book addresses the interior human experiences. A particularly large number of stanzas deal with how dangerous it is to consort with women and how to reduce one's lust. Now, since he was a monk, I suppose it is reasonable that he should say something on this topic. But huge sections of the book are devoted to this sole topic—lust was apparently far more important to him than the birds or trees that lived around him. Since he had some 1000 verses to work with in his treatise, I think it reasonable to suggest that he could have composed at least one verse about the need to protect his native Indian forests from clearance and the need to appreciate wild creatures that lived with him. Instead, it is as if they hardly exist. Did he not awaken to a dawn chorus of birds, or sleep with a night chorus of frogs, or see lions and elephants sleeping in his forest? To use a modern term, they have been 'ghosted'.

Shantideva is not alone. One sees a similar dismissive attitude in modern Buddhists too. Like Shantideva, they go to the forest to practice meditation, but otherwise the wild inhabitants that live there are mostly ignored. I have spoken to a lot of people who have done retreats in wild locations and rarely hear stories about the plants and animals that were already dwelling there.

So, here is the problem, as I see it. Buddhists have been thinking about human nature and non-harming for nearly two millennia. And Buddhist liturgy is filled with references to having compassion for all living beings. This does not identify all humans, but *all* living beings. So why then, did we have to wait for someone like William Hornaday, an American, to exclaim "We are weary of witnessing the greed, selfishness and cruelty of 'civilized' man toward the wild creatures of the earth. We are sick of tales of slaughter and pictures of carnage," the statement that opens this chapter? His generation led the way to the wildlife laws that protect the wildlife of North America, laws which are a model for other regions of the world. And, similarly, why did we have to wait for John Muir, born in Scotland, to advocate for the protection of wild areas, and lobby for the creation of national parks like Yellowstone? We lost nearly two millennia when this message could have been spread and acted upon.

I wish I could say it is getting better. So, now for some personal observations. I have spent a lot of time in the forest, watching wild creatures. I have also spent a considerable amount of time at Buddhist events, watching Buddhists. I can report that wild areas and wild species are rarely discussed at Buddhist events, except in the most superficial way. Indeed, the primary conservation interests are mostly how to recycle coffee grounds and plastics. Recycling is a virtuous activity, but it constantly surprises me that people so often think no further than their own rubbish. I have never heard anyone talk about rhinoceros, coral reefs, wetlands or sea turtles unless I bring the topic up first. I have heard a few rumblings about global warming, but inquiry soon reveals that this is mostly virtue signaling and still is mostly a result of concerns for humans. That is, the concern over rising sea levels is for the impact it will have on humans in cities, not on the egrets and turtles that live in coastal marshes.

The best I can say is that when I am surrounded by Buddhists, I am not subjected to gory hunting stories or bragging about the trophy animal that was shot on the last trip to Africa. Overall, I would say most of my Buddhist friends largely view the natural world with benign neglect, as a backdrop to the human drama. Their families, their urban lives, and their spiritual practices are in the foreground, and wild nature is not perceived as an active part of their humanity, nor relevant to their survival. Yes, that popular word 'ghosting' comes to mind again.

Here is the final contemporary example. It has taken decades of hard work to build a protected wilderness areas system in Nova Scotia. There is good news to report: a proper system is emerging. There is also now a Buddhist monastery, Gampo Abbey, snuggled neatly between two important wilderness areas: Cape Breton Highlands National Park and the Polletts Cove-Aspy Fault Wilderness Area. What an amazing location for a meditation retreat, but there is nothing about either of these wild places on the Gampo Abbey web site. There is also no mention of local wildlife, not even, say, the threatened Bicknell's thrush which nests in these adjoining wilderness areas. You will, however, find photos of some abbey residents holding a sign that says 'black lives matter'. Not thrushes, you note. Not Blanding's turtles. Not right whales. Not Canada warblers. Even though these Buddhists live in Nova Scotia, with important wilderness literally right next door, the retreatants' minds are dwelling instead on their own species, in cities far away.

Updating the Buddha's biology and the risk of speciesism

The Buddha lived more than 2,500 years ago, when the Earth's population was smaller and the technology for killing much less efficient. Shantideva himself lived more than a millennium ago. Perhaps it is time to update our relationship with wild nature. The world faces new and urgent problems. We can, of course, ignore them and focus on ourselves and on our families, friends, and pets. Or we can engage with the natural world and include this within our spiritual path. The decision to ignore the natural world, and to focus on ourselves, is a new kind of dangerous *ism*, and it deserves a name: *speciesism*. Speciesism says that we need to think only about our own species and maybe a few animals like our pets and farm animals, and the world exists just to feed us and to provide a green backdrop to the spiritual path. Speciesism is dangerously close to solipsism, and it may be an unexpected risk of meditation and of focusing upon manuscripts that are more than a thousand years old. If we move our attention inside, it is easy to forget that vastness also surrounds us. That vastness is filled with living beings. We have liturgy claiming that we are responsible for all living beings. If we simply ignore those millions of beings that comprise wild nature around us, Buddhists risk being called to account for blatant hypocrisy.

So what can Buddhists do to catch up? In a special section on the next two page spread, I suggest ten positive actions.

Looking around my own Buddhist communities, I see so little action that it is easy to get discouraged, if not annoyed. At least there is a positive angle: we have immediate opportunities to begin changing for the better. We can start today. And, since we are fond of calling ourselves practitioners, we can make these activities a part of our practice. More on this later in the book.

Out of sight, out of mind, out of luck? Many traditional Buddhist teachers have told their students to live in the forest. However, a forest is much more than a convenient green backdrop to human activity; it is a living community with its own inhabitants. Here is one example, the threatened Bicknell's thrush, which nests only in remote areas of eastern North America, including northern Cape Breton Island. The survival of such wild animals depends largely upon humans and whether we can open our hearts and minds to their existence.

Making places for wild nature in our minds and on our lands

As I said, conversations on the environment in my own community usually focus upon recycling. People can be quite intense on topics like the fate of yoghurt containers or coffee grounds. (Even people with tattoos.) Here I want to emphasize the urgent need for us to look far beyond the recycling bins of our life and turn our attention outward to wild landscapes.

Here is the good news. Conservation groups are already urgently working to buy critical wild habitats. Well-known examples include large groups such as *The Nature Conservancy* and *World Wildlife Fund*, but there are many other conservation groups and land trusts operating in local communities. Land is still for sale everywhere you look. Will it be bought for conservation or for logging and urban sprawl? I just checked online at the *Rainforest Trust* to confirm that you can still buy threatened rainforest in places like Borneo and Amazonia for just a few dollars an acre (rainforesttrust.org). At this price we can all participate. So buy some rainforest today. And tomorrow. And the next day too.

Other land, of course, costs more, but land trusts around the world are looking for important properties to acquire and raising the money to do so. The good news is that if we make the right decisions with our own money, we can have a huge positive impact on wild species. Of course, it helps when the government participates, but it is far too easy to avoid personal responsibility by expecting the government to do everything. It is too late for the giant ground sloth and the passenger pigeon but not too late for millions of other wild species that share the Earth with us. Even if you live in an urban apartment, you can connect with wild nature by supporting a land trust.

There is more good news. We now have a growing international network of protected areas. Yellowstone National Park, Kruger National Park, Great Smoky Mountains National Park, and the Great Barrier Reef Marine Park are just four out of hundreds of examples. We can enjoy our hiking trip through the Great Smoky Mountains only because past generations made the effort and paid the costs of protection. We now need to do our part and complete a global network. This will mean expanding existing protected areas and adding new ones. This is one of the moral imperatives of the century. It will take money. It will take popular support. It will also take leadership.

Positive actions everyone can do—today!

1. When we teach about loving-kindness or compassion, we could emphasize that these attitudes are to be extended not just to humans, but to nonhumans as well, and mention specific groups (e.g., World Wildlife Fund, The Nature Conservancy, National Audubon Society) already engaged in compassionate and positive actions. We don't need another talking circle.

2. We could expand our view of 'social engagement' beyond issues of gender, race, and poverty—solely human-centered concerns—to include engagement with our nonhuman neighbours and wild places.

3. We could surround our homes, schools, and retirement homes with habitat friendly to wild nature by reducing or eliminating mown lawn and replacing it with native trees, shrubs and ground cover including plants for pollinators.

4. We could build a personal relationship with a wild place near our home. This includes learning the names of the common species that live there, speaking on behalf of that wild place as necessary to ensure that it continues to be respected and protected.

5. When we think about personal purchases such as tattoos, sports event tickets, or travel, we could ask ourselves whether the money would be better spent buying land. For your holiday or birthday gifts, ask your family and friends to donate to organizations that buy and protect wild places.

6. When we plan our donations and bequests, we can give some of our money to groups that protect wild places and wild species. Does your will include a donation for rainforests or coral reefs, wolves and cougars? If not, then change it. Now.

7. When we teach about (or practice) compassion or right action, we could include actions such as buying wild land and protecting endangered species.

8. When we discuss sustainability or conservation policies in our centers, we could think beyond recycling and become friends of a local natural area (or adopt a local threatened species).

9. We could ensure that all our land centers and retreat centers have policies in place for protecting and enhancing natural habitats and that their web sites specifically describe the natural environment and adjoining wild areas.

10. When we are responsible for youth activities, or in raising our own children, we could take them out to see wild places, help them learn about their wild neighbours, and buy them binoculars, field guides, and camping equipment.

You are welcome to photocopy and share these two pages from *Darwin Meets the Buddha*, Paul Keddy, 2020, Sumeru Press, Ottawa.

Here is a final inspiring example. The many islands off the east coast of Nova Scotia, Canada offer stunning wilderness landscapes, and they provide habitat for many wild creatures. Within the last decade, the Nova Scotia Nature Trust has acquired not just one, but nearly 100 islands for protection in perpetuity. Their fundraising site calls them "The Most Valuable Islands You've Never Heard Of." The outpouring of financial support for this program says a great deal about the other side of human nature: our ongoing love affair with wild nature. We really do long for those days in the Garden of Eden. And we can still make it happen.

8
Getting Along:
An Ecological View of Compassion

> Do unto others as you would have them do unto you.
> *(Jesus, The Sermon on the Mount)*

HUMANS TEND TO THINK of life being made of organisms and individuals, with a particular emphasis upon the individual and the 'self'. It is sometimes helpful to look at the world from a different perspective—that of the gene. From the perspective of the gene, each individual is quite expendable. Although it is not, at first thought, a familiar or cheerful view of reality, it is what it is. What do we learn when we think of living organisms as being mostly factories for manufacturing new copies of genes? In this worldview, the underlying reasons for many kinds of human behaviour can be measured in one currency: the relative number of copies of genes that pass to the next generation. Hence the origin of the title for Richard Dawkin's book, *The Selfish Gene*, a book that provided important new insights into the process of evolution.

Humans as gene factories

Our initial assumption when we first read about selfish genes is that selfish genes must be expected to build an entirely self-centered world. It turns out we are wrong. Whether we consider our cells, our bodies, or our human societies, non-selfish behaviours are not only possible, they are almost inevitable.

Should individuals help one another? Too many people who are unfamiliar with evolution, natural selection, and biology make the simplistic assumption that selfish genes cannot lead to co-operation. Perhaps it is a failing of the English language that leads us to assume that selfishness and co-operation are mutually exclusive. If we take such a simplistic view, we create a caricature of evolution that justifies social Darwinism, a view that justifies the exploitation of the weak and poor by the powerful and rich. Even the most determined advocates of selfish gene theory admit that there are compelling reasons for genes to build factories that care for one another.

We can even predict how much one factory should be willing to help another. The care we should lavish on those around us and the resources we allocate to our family and friends would be expected, on average, to increase with our genetic similarity. That is because genes can multiply not only by building individual gene factories, but by helping other factories, so long as they are producing copies of the same gene. In this case, there is a precise arithmetic of when an individual should sacrifice his life for another: an identical twin, for example, would be worth dying for, whereas an adopted brother would not. We must appreciate that this sort of arithmetic is not something we do consciously; indeed, it is only the last few years that ecologists and evolutionary biologists themselves have been able to work out the approximate rules. Rather, this process has been going on for millions and perhaps billions of years, as organisms with genes that happened to make the right decision proliferated at the expense of those that did not. A gene, for example, that encourages an organism to die for any neighbour, a complete altruist, would be unlikely to survive when competing with genes that acted selfishly. An obvious and most illuminating exception of sorts is found in the social insects. Ants and termites freely die to protect the colony. Although they look to us humans like they are individuals, they are not, in the human sense. All the workers are genetically identical.

While we may find it distasteful, the tendency of genes to prefer to co-operate with genetically-similar entities is self-evident in the natural world: most parents feed their own children rather than a neighbour's children, most parents leave their money to their children rather than to complete strangers. If you read the obituary columns in newspapers, you will note too, that many obituaries proudly list the number of children and grandchildren produced. There are exceptions, of course. But when parents kill their children, or parents leave their money to complete strangers, the fact that it is considered newsworthy proves how uncommon such behaviour is. Such incidents are newsworthy because they challenge the common way the world works.

The idea of co-operation even provides a possible explanation for behaviours such as homosexuality. There are many examples of co-operation within family lineages. Some individual birds, for example, may not breed but instead remain in the parental territory and assist their parents in raising new young. In such circumstances, while those certain individuals do not reproduce directly, they contribute to the spread of family genes by assisting in the gathering of resources and care for family members who carry the same genes. In the extreme case, such as social insects, a vast majority of the individuals forego reproduction entirely and instead care for a small number of reproductive individuals. But the social insects are a most unusual case and have been the object of entire books themselves. They are an extreme example of the principal that individuals can increase their gene transmission without directly making offspring themselves.

A brief return to conflict

Before we proceed further with the topic of co-operation by genes, we should digress and admit that selfish genes can indeed cause harmful behaviour. Harmful, at least from the perspective of individuals. It is entirely possible for teams of people to co-operate to exterminate other gene lineages. The underlying lesson here, however, is the degree to which this human behaviour is driven by the principle of relatedness.

Consider this aspect of relatedness in the Old Testament. In order to increase the number of genes one leaves, it is sometimes necessary to take resources from others and to remove their capacity to leave offspring. The Hebrews were in constant conflict as they tried to wrest land away from the Canaanites, who already occupied the area, and as they encountered the Philistines, who were already settling along the Gaza coast. In Deuteronomy 20:13 and 14, God gives specific instructions for competition with these peoples. Talking about besieged cities, He commands "And when the Lord thy God hath delivered it into thine hands, thou shall smite every male thereof with the edge of the sword: But the women, and the little ones, and all that is in the city ... shalt thou take unto thyself." However, in certain nearer cities which God has given the Israelites "...thou shalt save alive nothing that breatheth: but thou shall utterly destroy them;" (Deut. 20:16-17). In summary, God advises, extirpate your nearest competitors, but keep alive and enslave the females of the slightly-more-distant neighbours. While the men are killed, the females are kept for reproduction. The result will be that your genes will spread in number relative to theirs.

Nor is it just the Bible. Consider the massacre in the Market of Medina in 627. This account is simplified from Andrew Bostom's recent book *The Legacy of Jihad: Islamic Holy War and the Fate of Non-Muslims*.

The Jewish tribe called the Qurayzah had surrendered to Mohammed. After due deliberation, a ruling was reached. The men (somewhere between 600 and 900 of them) were separated, trenches were dug, and the men were decapitated and thrown into the trenches. The women and children were sold into slavery, a number of them being reserved for Mohammed's companions. According to one biographer, Mohammed chose a Qurayzah woman named Rayhana for himself. The property of the now-extinguished tribe was divided up among the perpetrators. This incident and its role as a precedent have been extensively discussed by commentators through history, including in the aforementioned book.

In such cases, one gene lineage is ended, and resources are directed into a new lineage. Yes, we have to admit that selfish genes can indeed produce harmful behaviours. Again, the point is not that humans consciously calculate the genetic outcome of such behaviour. It is instinctive, working at the level of the subconscious or even the unconscious. The voices of various gods that justify such behaviours may indeed be coming from our own genes.

Although we live in different times, our behaviour evolved in times when we lived as tribes or large family units. In those ancestral circumstances, it was fairly obvious who qualified as 'us' and who qualified as 'them'. 'They' are in another tribe, perhaps in another valley and may even have a different god. Xenophobia, the fear of outsiders, can be seen to have a possible genetic basis: 'others' are those who are strange enough that they are unlikely to closely share genes with us.

Co-operation and competition

Having done due diligence by addressing some negative effects of selfish genes, let us get back to the topic of co-operation. For one essay on selfish genes, Richard Dawkins chose the title "The Selfish Co-operator." He has been insistent that we should not draw the wrong conclusions from the scientific understanding of selfish genes. Selfish genes can and do produce co-operative and even compassionate behaviour.

Consider that much of our behaviour evolved when we lived in small groups and tribes. Intermarriage meant that most or even all of those around us shared a common genetic heritage. If we looked after one another, like providing extra meat from a hunt, we indirectly assured our genetic survival and propagation. When living in a small tribe, fine calculations about degrees of relatedness were probably less necessary. Indeed, one could even take the argument a step further. Too much awareness of and worry over genetic distance could create dissension within the tribe and lower everyone's chances of survival. Suspicious and argumentative tribes may have self-destructed. Again, the evolutionary arithmetic: perhaps it is better to tolerate the occasional doubt about parentage than risk conflicts that would destroy the tribal cohesion that will advance the survival of one's grandchildren.

Compassionate behaviour has evolutionary value

Let us return to first principles and consider some of the many situations where co-operation and compassionate behaviours would have been of value to our ancestors. There is a large scientific literature, and many books focus on this topic. Here let us keep it simple. You can read Dawkins and the literature on kin selection later if you wish.

The care of offspring is found in many animal groups including birds, amphibians, reptiles, fish, and insects but is most thoroughly developed in mammals. Just as offspring are conditioned to revere their parents, parents must love their offspring. Love and affection produce a set of behaviours that includes the desire to protect and feed one's young. It also conveniently reduces the probability that a hungry parent will eat its own young.

The care of one's own offspring might lead to a similar response to the offspring of neighbours. This might be entirely accidental. It might be a mere side-effect of being attracted to small, helpless creatures with large eyes, like dogs and puppies. When we hear "Aw, isn't it cute," we are actually hearing from that set of genes that drives us to protect infants. If we care for a neighbour's offspring rather than our own, it might just be that we are applying this human behaviour in situations where it has few benefits, but also few costs. But again, returning to ancient tribal life, a neighbour's offspring may very well carry some of our genes.

If one shared resources with a less-fortunate neighbour during periods of good fortune, this could be expected to increase the probability that a neighbour would reciprocate. Sharing would then be favoured as a means to enhance survival during crises, a situation for which we now increasingly depend upon insurance and pensions. Neither existed when humans were evolving, and the care of the tribe was all one had in the case of illness or injury. Thus, it is entirely possible that generosity too could evolve in a process that is driven by selfish genes. Anyone who has lived in rural communities where bartering and helping are a normal kind of behaviour will likely be struck by the complexity of these helping behaviours, the degree to which people can remember and balance out their assistance to one another (and the degree to which they can remember and act upon grudges when such behaviour is in some way violated).

Sharing resources may have enhanced one's social status, so, whenever surplus resources were available, there may have been social advantages to co-operation. We have seen how prevalent hierarchies are in human life, and how they are often linked to aggression and the exercise of power. Compassion could be thought of as an antidote to aggression, a psychological state that enables tribal units to survive and multiply. Sharing might even be a means of climbing the hierarchy. In this sense, compassion is not unlike the biochemical signals that allow groups of cells to co-operate with one another within individual bodies. Each cell has to co-operate with many other cells if the community of cells is to survive.

In summary, the need to care for young and nurture relationships with other people was probably a powerful incentive for developing the sense of caring for one another in general. Compassion, love, kindness, caring, and self-sacrifice may be just as encoded in our genes as physical characteristics such as opposable thumbs and binocular vision.

Compassion and social organization

So, compassion can have entirely Darwinian origins. Compassion, in the Buddhist tradition, is an essential part of the path to enlightenment. Compassion is the topic of many a lengthy Buddhist treatise. Compassion is regarded as one of the Four Immeasurables, one of the four qualities of an enlightened person. Compassion, then, is not just the ability to pity another unfortunate person, but rather has a larger set of meanings. Compassion has the quality of an expanded, aware, gentle heart that sees the true human condition: we are alone, we are afraid, we suffer; and the other beings around us feel the same way.

Societies still strive to find a balance between rights and duties, between the selfish good of the individual and the benefits to all from social harmony. We could propose that ideologies such as socialism are a political manifestation of instincts for compassion. From this perspective, social Darwinism—the empowerment of the rich with complete discard for the poor—is literally inhuman. That is, to produce a society that allows the expression of aggression, but not compassion, is as unenlightened as if we produced an environment that strengthens our arms but allows our legs to wither away. Yes, there are parts of us that are selfish, but equally there are parts of us that feel compassion and the urge to co-operate.

>Love thy neighbour as thyself.
>Do unto others as you would have them do unto you.

These sentiments may not be the naïve hopes of a few religion-intoxicated prophets. They may be our evolutionary heritage and the highest expression of our social consciousness. Although these sentiments may have originated in the tribe to promote social cohesion and the survival of our offspring, they are no less vital than our lungs and our hands. Just as we exercise our bodies and educate our minds, compassion is a human attribute that we can deepen and cultivate. There are specific practices for this purpose, such as visualization of compassionate deities. According to the Buddhist teachings, as we cultivate these psychological states, they actually dissolve many of the painful states that cause us suffering. It seems that we are in fact designed to feel pleasure when we are helpful to others and pain when we are selfish. This pleasure provides a space in which we feel useful, valued, and worthwhile as members of our community. At times when we despair of the human condition, we can remind ourselves that our evolutionary past has equipped us splendidly for social cohesion.

From this perspective, the teachings of Jesus Christ may have had their broad appeal precisely because they resonated with deeply held human qualities that were inadequately expressed in the world at the time he taught. The progression from the Old Testament to the New Testament can be seen as an attempt to adjust a set of religious teachings to a closer concordance with the inner longings of humanity. The Old Testament fits with those parts of the psyche that desire to appease a parent, maintain power within a social hierarchy, and accept rules created by the tribe, but there is perhaps a starkness created by lack of love, by insufficient compassion. The longing for something more than mere obedience is perhaps best illustrated in the Psalms and the pleas of early prophets. The voice crying in the wilderness may have been a voice crying for a set of teachings more in accord with the innate human mind, a view of the world that valued and empowered the kind, caring, compassionate side of humanity.

The dark side of human existence is also discussed in the Old Testament and well-illustrated by the tragedies which befall Job. When, his fortune gone, his family dead, his body covered in boils, he loses patience and finally cries out to God at the unfairness of life, God's only answer is that of a tyrant. God goes on for many verses bragging about his power 'Knowest thou the ordinances of heaven?... Canst thou send lightnings?," and Job is expected to gird up his loins and take it like a man. This is a far-from-satisfactory explanation for cruelty and an equally unsatisfactory prescription for living with it. Consider, in contrast, the New Testament

> ...though I have all faith, so that I could move mountains, and have not charity, I am nothing. (I Cor. 13:2)
>
> And now abideth faith, hope, love, these three; but the greatest of these is charity. (I Cor. 13:13)
>
> Blessed are the merciful: for they shall obtain mercy. Blessed are the pure in heart; for they shall see God. Blessed are the peacemakers: for they shall be called children of God. (Matt. 5: 7-9)

The qualities being described here come very close to those taught by the Buddha. Generosity is the virtue that produces peace. Compassion is the path to liberation. Egolessness provides an escape from suffering. From the oceans of suffering may I free all beings. It is possible that teachings such as these have survived and even proliferated because they deeply accord with some of our most basic longings as human beings.

A cautionary note on idiot compassion

Overall, the good news is that compassion is a natural part of our human inheritance, even in a world driven by natural selection. At very least, compassion exists to minimize the harm we cause to one another. Also, it can inspire us to engage in positive actions that will change our world for the better. Meditation can help us cultivate our compassion, and we will see in Chapter 10 that certain styles of meditation can particularly arouse this emotional state. Meanwhile, while we are following the Eightfold Path, we notice compassion when it arises, and we cultivate it.

All the same, some words of warning are necessary. Just because we feel compassion in a particular situation, the mere fact that we are feeling that emotional state does not guarantee that our responses will necessarily be beneficial. That is, we have to act with skill, with discernment. If we act on our feeling of compassion without consideration of cause and effect, our actions can create further harm. One well-known example is the phenomenon of 'enablers.' Family abuse and drug addition, for example, can be perpetuated by such people. The enabler may indeed be feeling real compassion for someone who is in pain, but their actions, in response, actually create further suffering. Inappropriate actions in response to feelings of compassion have been called 'idiot compassion.' Although there are many possible examples of idiot compassion within human societies, let us consider, instead, some examples that are likely to be less familiar—idiot compassion in regard to wild nature.

It is not that long ago that some well-intentioned people who saw declining wildlife populations felt intense compassion for selected species, such as ducks and deer. These same people therefore shot, trapped, and poisoned predators including hawks, eagles, wolves, and cougars. It may seem obscene to us today, but it happened. Even today, poisoned baits are still killing wildlife. That is one reason why we will not likely hear wolves or see cougar tracks on this hiking trip. But we will see lots of evidence of deer. Way too much evidence, in fact, which brings us to our next example of idiot compassion. For some decades we have had an enormous problem with overpopulation of deer, partly caused by the deliberate extermination of their natural predators, principally wolves and cougars. Now we reap the whirlwind: deer are destroying the ecological integrity of our remaining forests, even in large protected areas like Great Smoky Mountains. Not only do the deer kill many of the native wild plants in our forests, such as spring flowers and wild orchids, and not only do the same deer kill the young trees that should eventually replace the older trees, but in some cases, deer have even been observed eating the young in the nests of ground-nesting birds! In fact, my own forest is suffering heavily. Wild plants like Canada yew and hobblebush have vanished from my property within my lifetime. Trees like eastern hemlock cannot re-establish. Yet, some of my neighbours insist on feeding deer during the winter because these neighbours like to feel good about feeding a hungry animal. So, the neighbour feels good for a few hours and the deer survive the winter, have even more fawns, and do even more damage to the forest; then next year there are even more deer to be fed. It is a vicious cycle that eventually will lead to the degradation of the very forests that provide winter shelter for the deer themselves. Idiot compassion has unhappy consequences.

This point is important enough that it deserves repetition. The feeling of compassion, can, if we are careless, lead us to cause more harm. That is why we call it idiot compassion. Let me close with one more example from a wild forest. One of our flagship protected areas in Ontario is Algonquin Provincial Park. It has its own population of wolves, huge areas of deciduous forest, and many significant plants and animals. Yet the park is still being logged and is criss-crossed by a network of logging roads—some 8,000 km of roads! The usual justification for this sad situation uses compassion: we are told that logging the park creates jobs for unfortunate, poorly-educated people in nearby towns. So, the logic goes, compassion for humans requires us to sacrifice our native forests and many of the wild creatures that live there. This same compassion requires us to look the other way as logging trucks drive over migrating frogs, salamanders, and turtles. Something is wrong. My point is that many cruel or harmful acts can be camouflaged by some sort of pseudo-compassion. It is sad that people are poor, but cutting down the forests in our remaining wild areas is not a real solution to rural poverty.

Furthermore, we know from history that even tyrants will often cloak their harmful behaviour in pseudo-compassion by promising a solution to a problem that is causing discomfort. You could make a short list from history of these sources of discomfort. Three common ones are: fear of disorder, fear of unemployment, and fear of other tyrants who are even worse. A good tyrant will promise to help us remove one of these from our lives, even if it takes a military coup. Sadly, it seems that compassion can be used as justification for harmful acts. We want to feel good about ourselves, and sometimes superficial activity makes us feel just as good as skillful activity. One supposes those early game wardens who exterminated wolves and cougars in wild places felt they were engaged in a virtuous activity, which is why old black and white photos show them posing proudly, even virtuously, with hanging wolf carcasses.

All this is not to suggest in any way that compassion is not an important attitude. It is inherent in the Buddhist path, it is foundational for any human society, and it is buried in our ape instincts too. Yet it is vital that we join our feelings of compassion with our rational capacities. If there were space in this book and time on our trip, we could have a lengthy discussion of idiot compassion with more examples from the modern world. Alas, we will have to set these topics aside for another occasion, perhaps a journey by canoe through Algonquin Provincial Park, where we are reliably assured the logging companies show their compassion by avoiding the use of noisy chainsaws along canoe routes during the summer visitor season.

Building on compassion

How do we build a modern world where our inherent quality of compassion is valued, cultivated, and skillfully applied? We have seen how easy it is to lose track of compassion and end up with various kinds of horror. Perhaps we could begin by revisiting traditional civic virtues: duty, self-control, community participation, and public service. The early Greek city-states did not understand evolutionary psychology, but they managed to create a realm in which citizens had a measure of control over their society and where early democracy was able to take root. Similarly, several millennia later, the American Constitution enshrined the right to life, liberty, and the pursuit of happiness. Although happiness today is a somewhat shallow word, at the time it had a meaning far closer to liberation, enlightenment, or self-fulfillment. Duty, self-control, community participation, public service, and self-sacrifice are all names we give to particular qualities of co-operation and compassion. They may arise from our need to create and nurture a compassionate, or even enlightened, social environment.

Individual humans can use their compassion to transform their society. One of Canada's most-loved politicians was Tommy Douglas. He is credited with creating some compassionate government policies that are now central to the Canadian identity: unemployment insurance, old age pensions, and free medical care for all citizens. Younger readers take these for granted, but they did not exist in his childhood, which was marked by hardship, hard work, and personal illness. His biographer, Vincent Lam, describes how Douglas was shaped by the suffering he saw on the Canadian prairies during the Great Depression. In 1919 he witnessed a dark incident in Canadian history: June 21, 1919, the day armed and mounted policemen and an extra team of "special constables" used guns and clubs to suppress the Winnipeg General Strike. Douglas was inspired to work for social change as an "expression of God's will for men on earth" and became a small-town minister and then eventually moved into politics. Throughout his career he was motivated by the teachings of Jesus. He asked his audiences, how can we make practical changes that create a world that is more caring for our fellow human beings? Of course, many other people worked with him to bring these constructive changes to Canadian society. And these changes came not through violent revolution, but through the ballot box and through our parliament. Tommy Douglas gave his last speech in 1983 to the New Democratic Party, at the age of 79, weakened by cancer. His biographer reports that when he finished, the cheering started, and it is said, lasted for a full 23 minutes, an outpouring of "joy, gratitude, and admiration."

Values like civic duty, self-sacrifice, and compassion, then, are not naïve qualities or wishful thinking. They are part of our humanity. Compassion is part of our evolutionary heritage.

9

A Brief History of Life:
Co-operation and Community

Up to this point on our hiking trip with Darwin and the Buddha, the conversation has mostly explored the behaviour of humans and the experience of being human at the level of the individual. We've also talked about how that individual relates to other individuals and how an individual carries with it a somewhat imaginary 'self'. That is part of our evolutionary baggage that sometimes seems to weigh heavily upon us. The feeling of being an individual is a very strong one, yet when we hike along a trail, are we not also a team? We share information, swap food, and help each other pitch tents and adjust packs. Even our dialogue is a group venture. So, let us look more carefully at how our own, apparently-individual body is in fact a similar team project.

In some ways, we are not an individual at all, but a colony. It is a simple biological fact that our bodies are built out of cells. Even our ape minds are built of cells. Our bodies are also hosts to nearly as many bacteria as cells. What, then, is an individual? Let us consider some possible lessons from the nature of our own bodies. This requires us to look back into the past, nearly four billion years into the past.

How did existence come to be? The Buddha was apparently quite unwilling to get involved in discussions of how the universe came to be. He often, it is said, made it clear to questioners that his teachings were not mere idle speculations. They were focused on what *is*: the First Noble Truth and the Eightfold Path. He used the example of a human injured by an arrow. It is a waste of time for this person to worry about who made the arrow, what kind of bird furnished the feathers, or where the wood came from for the shaft. He needs to visit a doctor and have the arrow removed. So the Buddha taught.

Of course, the Buddha taught in a time when there were no microscopes or telescopes, no written reports of scientific expeditions like *The Voyage of the Beagle*, or satellite images of the Earth's forests. Fortunately for modern humans, we know a good deal more about our planet, its life, and the origins of life. Instead of speculation, we now have facts, from the age of the Earth to the ages at which various kinds of life arose. Of course there will always be room for refinement, but the story is now quite well-understood. A really good overview of the history of life can be found in *The Earth Through Time*, by Harold Levin. Since it has been through more than ten editions, he must be doing something right! It has some nice drawings of dinosaurs, of course. But here I mostly want to talk about life long before it included large terrestrial animals. I will give you a short overview of two very early stages of the story of life, chosen because they illustrate how co-operation is also an essential part of the story of life.

Modern science therefore allows us to give a certain kind of answer to Job, the figure in the Old Testament, who, in his suffering, represents all human beings. Job's experience is similar to our own: we find ourselves alive, we suffer, and we do not know why. Poor Job had multiple problems including the death of his children and loss of his wealth, compounded finally by a painful case of boils "from the sole of his foot to the crown." When Job complained to God about this misery and demanded to know why it was happening, God's response was an insult. God's response in the Bible mostly says that Job is so ignorant and weak that God does not need not provide him with any explanation for suffering.

Job was poor, sick, and untutored in science. So he had to admit that he did not know the answers to God's questions. Today we might be able to tell Job a little more about how he ended up in his predicament. It would go something like this.

Survival of the fittest can produce co-operation

Darwin is widely associated in the public mind with the phrase 'survival of the fittest'. Just what is meant by being fit? In the broad sense it means surviving and producing offspring and, more precisely, making more copies of one's genes. In fact, the term 'survival of the fittest' originated not with Charles Darwin himself but with Herbert Spencer, who was trying to find another way to explain Darwin's term of 'natural selection'. Darwin then picked up Spencer's usage and used survival of the fittest later in other books. He did so in part because Alfred Wallace, the co-founder of evolution, wrote him and recommended Spencer's writing. At that time, then, many people were reading Darwin (and Wallace, Spencer, and many others) as they tried to develop the correct words for describing the process of evolution by natural section. Today, more than a century later, the common understanding is a kind of caricature, reducing it to the idea that fitness involves the ability to compete with other individuals. Hence the phrase, nature 'red in tooth and claw', that actually comes from the poet Tennyson.

To be clear, the phrase 'survival of the fittest' is an awfully good starting point for understanding biological reality. We are here only because our ancestors were able to gather resources, survive, and reproduce, and this reality takes us back not only to the first humans in Africa, but back to the time life first emerged on land and back even further to the first multicellular organism to appear in the ocean. They are all our ancestors, and we are alive only because they survived and reproduced. So, survival of the fittest also succinctly cuts through a lot of wishful thinking about the human condition. I am not referring just to the wishful thinking that we are specially created in the image of God, but wishful thinking all together. It is wishful thinking to imagine that we can burn the world's coal reserves without changing the climate. It is wishful thinking that we can allow the human population to continue expanding without harmful consequences for everyone. It is wishful thinking to imagine that we can solve overpopulation in Africa by importing the excess population and giving them free houses and medical care in Canada. It is wishful thinking to imagine that wild species, particularly large animals and particularly predators, will continue to exist unless we set aside more large wild areas. Wishful thinking and willful blindness are serious problems and really do contribute to samsara and dukkha. So that somewhat loaded term 'survival of the fittest' does have its merits: it demands that we think accurately about biology and natural resources, as we have done in some preceding chapters.

Having said that, we also have to beware of oversimplifying our biological reality. It is entirely possible that fitness requires a good deal of co-operation. A single tree, for example, may depend upon fungi to collect soil nutrients, insects to pollinate its flowers, and mammals to disperse its seeds. A forest therefore has a great many co-operative interactions. (If you are looking for more examples, note that ecologists tend to use the world 'mutualism' for co-operative interactions.) So, in this chapter let us look more at biology from the point of view of co-operation. Let us focus, in particular, on our bodies and their origins. Here, in advance, is the conclusion: our bodies, at multiple levels, are the result of natural selection acting through co-operation.

My intention is to walk you through a couple of phases in the origin of life, which is a classic, but often overlooked, story in biology. I will focus in particular on two phases where co-operation turns out to have been vital: the origin of eukaryotic cells (those are the kind of cells that comprise much of our human body) and the origin of multicellular organisms (that is, groups of cells also occurring as part of our human body). It is likely unrealistic to assume that one chapter can convey four billion years of evolution but, nonetheless, it is part of our heritage and something about which we should know at least a little.

The foundations of life

Life appears to require a situation near the edge of a star system where the energy being produced is radiating off into outer space. We do not yet know how often such conditions occur; it may not be infrequent, given that our own galaxy has some 100 billion suns, and a good many of those likely have planets. We know that some of these stars have their own solar systems, and many more likely occur but are too small to be seen. Our particular planet is located in such a way that we have energy from our sun passing through our planet and dissipating in outer space. If we are looking for the origin of our own life or other forms of life, we need to find zones where energy is flowing through the cosmos. Life, at least as we understand it, is matter that has been organized by energy flow.

The basic work on this topic can be found in a small but important book by Harold Morowitz, published in 1968, *Energy Flow in Biology: Biological Organization as a Problem in Thermal Physics*. Regrettably, one rarely sees the book cited. It is not an easy read and has some daunting equations. Morowitz asserted that we have to look at the relationship between physical laws and biological systems. He showed, using the basic principles of thermodynamics, how the simple process of energy flow is able to create complexity out of simplicity. Books like *The Vital Question: Energy, Evolution, and the Origins of Complex Life*, by Nick Lane, provide an updated and more popularized overview of the field, but realistically, you cannot get away from having to know some thermodynamics and some chemistry. The important point is that energy flow occurs on Earth and has the capacity to both create and organize organic matter. So our bodies are using processes that already occur in the natural world.

Building life from atmospheric raw materials

To start off the story of life on Earth, we have to begin by asking what the early raw materials were—what would have existed in the early atmosphere of Earth before there was life? We already know from Chapter 1 that our bodies consist of only a few elements that could have been common on the Earth at that time. But in what form did C, H, N, O, P, and S exist then, and how did they end up being assembled? That requires some detective work and is one reason why scientists are interested in comparing the atmospheres of different planets. It is possible that the four inner planets with their stony crusts once had atmospheres more like the present day gas giants of Jupiter and Saturn, with atmospheres containing mostly hydrogen and helium with trace amounts of methane, water vapour, and ammonia. Another source of evidence is the composition of volcanic out-gassings on present day Earth, since early volcanoes would also have contributed to the atmosphere. In all likelihood, the principle raw materials included water, carbon dioxide, and ammonia. If this mixture seems unfamiliar, there is a reason: the atmosphere today is rather different, being largely nitrogen and oxygen. The basic elements that were present in the early atmosphere and in the oceans became the construction materials for life. Water and carbon dioxide, for example, are key elements in the most common biological molecule on Earth. What is this most common molecule, you ask? It is cellulose, the major component of plants and the most abundant (by mass) molecule in the biosphere.

The Earth is about 4.5 billion years old, and various sources of evidence suggest that life arose quite early in the Earth's history, some four billion years ago. What sorts of physical conditions likely existed back then? Other than sunlight, there would likely have been energy discharges such as severe lightning storms and energy gradients associated with thermal vents in the oceans. With energy flow and simple mixtures of gases, increasingly complex molecules were formed. For example, take a mixture of gases such as carbon dioxide, hydrogen, nitrogen, and oxygen, put them under 500°C, and the mixture yields mostly water and carbon dioxide with smaller amounts of other molecules, methane and ethane, which have higher potential energy. The latter two are less likely to form because these larger, more complicated molecules require more energy to create. As energy is pumped through the system, the distribution starts to shift upward toward more and more complicated molecules. Morowitz postulates that energy flow through the early atmosphere yielded similar results:

starting off with simple low-energy molecules like water, carbon dioxide, and nitrogen, more complex molecules were produced. The production was driven by an external energy source, which on Earth is our Sun.

As energy flow produces more complex molecules, there will, however, be a reverse tendency for these to fall apart again into simpler molecules. Some molecules, however, will be more stable than others and will tend to persist and accumulate. Others will fall apart rapidly. Obviously, those that are stable will accumulate relative to those that are unstable. Once there is a pool of larger molecules like ammonia and methane, they, in turn, can start interacting with each other, producing molecules with greater complexity and higher levels of potential energy. These too will have varying degrees of stability and will tend either to accumulate or fall apart. This process continues with increasingly complex molecules forming as a consequence of this external energy flow, the unstable ones falling apart, the stable ones accumulating. Thus there is natural selection for stability and for persistence even at the molecular level.

Experimental work nicely complements the work based upon these thermodynamic calculations. In 1953, Miller and Urey set up a simple atmospheric system with a hydrological cycle sealed inside glass tubing. Water evaporated and then cooled and condensed, only to be heated and evaporated again. Electrical sparks simulated lightning. Primitive amino acids formed. The Miller/Urey experiment was simple and the first of its kind. It stimulated researchers to modify the experiment in countless ways, changing the raw materials or changing the energy source, allowing them to explore how early life might have progressed.

Once such molecules form, some process is needed to allow these complex molecules to persist and increase in concentration. This process would likely have had several components, (1) protective wall development, (2) the direct use of sources of energy such as sunlight, (3) the ability to form larger multicellular aggregations to buffer against short-term periods of unsuitable conditions, and ... yes, even (4) consciousness. That appears to have risen rather late in evolutionary history, and so we are getting ahead of ourselves by perhaps a billion years or so. Consciousness can be thought of as the ability to develop predictive models for future events, to both seek out resources and avoid trouble by taking evasive action. For example, if an organism knows that certain conditions are likely to bring winter, then it can store up food or migrate to a better environment.

Let us try to imagine the circumstances on Earth some four billion years ago, when God tells Job that He (that is God) was "laying the foundations of the earth." Pools of increasingly complex molecules are accumulating. Molecules that are stable will accumulate; those that are unstable fall apart, by definition. Stability is one key trait and the other is replication. Any molecule that tends to create copies of itself will accumulate more rapidly than other molecules. Richard Dawkins suggests that 'replicators' are the key event in the origin of life as we know it. A replicator creates copies of itself, and people are still doing that today, several billion years later. If we think about the very early forms of life, chemical stability is what we now call 'survival', replication is what we now call 'reproduction', so really, even very basic molecular systems can develop the processes that we find in whole organisms.

Much of this story of life may be already familiar to you if you went to a reasonably good high school. There is even an episode of *Star Trek: The Next Generation* where Captain Picard goes back in time to visit early Earth and sees a 'little pond of goo'. The exact process by which early cells came about and where this occurred are, of course, still being studied. There are many unanswered questions. It was a long time ago, well before the modern fossil record. According to Nick Lane in *The Vital Question*, the oldest rocks yet found on Earth (at Isua and Akilia in southwest Greenland) are 3.8 billion years old and already show chemical evidence of organic life. Most of us were taught that the assembly of early cells likely occurred in shallow water, and we may visualize shorelines and a little pond of goo. Nick Lane, on the other hand, argues it is more likely that early cells originated under the ocean where thermal and chemical gradients were intense. He offers an account of early membranes and cells arising in deep sea thermal vents. And not just any kind of deep sea thermal vents, but alkaline thermal vents. These he says "provide exactly the conditions required for the origin of life: a high flux of carbon and energy that is physically channeled over inorganic catalysts" (p. 109). I can't explain any more of the chemistry since I don't pretend to understand it. But getting back to the topic, he says, "Thermal currents through microporous labyrinths have a remarkable capacity to concentrate organic molecules." So, it is just possible that deep within each of us is a biochemistry based upon hydrothermal vents like the ones found today near the Mid-Atlantic Ridge in the Atlantic Ocean. It all sounds quite romantic that these physical and chemical processes are still occurring inside each of us today.

The membranes that formed around each cell may have arisen to control the proton gradients that occur in such locations. Without a membrane one can argue that there really is no life. It is a membrane that divides the world for the first time into an inside (a living organism) and an outside (the organism's environment). When and how might membranes have first appeared? It was certainly early in the history of life, since nearly all organisms share the common structure of a membrane, a lipid bilayer. All of our bodies contain similar membranes. Another group of ancient organisms, the *archaea*, however, do have their own style of membrane, which seems to indicate that they diverged from other life forms early in the cellular history of life.

Eukaryotic cells are a co-operative venture

Although we now accept that all living organisms are built from cells, that knowledge is only a few centuries old. The discovery of cells in 1665 is attributed to Robert Hooke, who used a simple microscope. A few years later, in 1674, another biologist, Anton van Leeuwenhoek saw the first living cell. Darwin knew about cells; the Buddha did not. Microscopes are a recent invention. And now let us move on to the topic of co-operation.

Cells have existed for nearly four billion years. There are actually two very different kinds of cells on Earth: there are small prokaryotic cells and the much larger eukaryotic cells. How much larger? A typical eukaryotic cell is somewhere between 10,000 and 100,000 times the volume of a prokaryotic cell. To put this in context, a typical blue whale is only 2,500 times the volume of a human being. For much of the Earth's history, prokaryotes were the principle type of life. They are small, relatively simple, and have a wider range of chemical processes that allow them to occupy extreme habitats which would kill eukaryotic cells. Higher plants and animals, including humans, however, are constructed from large numbers of eukaryotic cells.

How many eukaryotic cells are there in the human body? Needless to say, you can't sit down and count them with a microscope. And it depends on whether you include red blood cells, which comprise the largest number of cells in the body by far. There are some 25×10^{12} of them. That is, 25,000,000,000,000. But red blood cells no longer have nuclei, so they are somewhat of an exception. According to a team at the Weizmann Institute of Science in Israel, the total number of eukaryotic cells in the human body, including those red blood cells, is 3.0×10^{32}. Those are big numbers and hard for us to visualize in any case. But it means that our bodies are not an individual but a community. These cells must co-operate in order for us to carry a backpack, turn a page, gather the light from this page, transmit impulses to the brain, and interpret the words.

The same team of scientists tried to calculate how many non-eukaryotic (non-human) cells, mostly prokaryotes, are also living in our human body. They concluded there were 3.9×10^{13} which means, roughly speaking, for each human cell there is also a prokaryotic cell. We are nearly outnumbered by other cells within our own bodies. We are therefore not just a colony of human cells, but an ecosystem of human and non-human cells. Getting back to the title of the chapter, that is a whole lot of co-operation going on.

We still have not talked about the origin of our eukaryotic cells. Not only are they larger than prokaryotes, they have a much more complicated interior, with various structures like the nucleus and mitochondria that you may remember learning about in high school. This raises the apparently thorny problem: how did this leap of size and complexity occur? Here is where the topic of co-operation becomes important in evolution.

Much of the credit for our understanding of the origin of eukaryotes can be traced back to Lynn Margulis, who in 1970 proposed that eukaryotes are in fact symbiotic associations of several prokaryotes (the serial endosymbiosis theory). An American physician, J.E. Wallin, had raised this idea in the 1920s, and he even published a book entitled *Symbioticism and the Origin of Species*, but his enthusiasm for the concept was far ahead of the quality of his data. With the idea now resurrected, it is possible that at least three structures in the eukaryotic cell are thought to have a symbiotic origin; the mitochondria, the chloroplasts, and the flagella. If an early prokaryote was invaded by a bacterium, an alga, and a spirochete, the result would be rather similar to a modern eukaryotic cell. There is a good deal of evidence accumulating to support this view. Certainly, the mitochondria in our cells were once prokaryotes. The same is true of the chloroplasts in plant cells. As for cilia and flagella, there is disagreement, and we will leave that for another day and more evidence. Flagella, for example, have modern tasks as different as moving sperm cells and gathering light within eyes, and indeed may have arisen from a third class of prokaryotic life forms. The structure of the flagellum seems to be constant throughout the plant and animal kingdoms, from sperm to retinal cells. Even the flagellum in green algae and a human sperm are remarkably similar, which suggests they all had a common origin deep back in time.

Here is the important point. The mitochondria, in which energy transformation occurs within each cell of our body, originated as an ancestral bacterium that invaded a larger cell. A single heart muscle cell, for example, contains about 5000 mitochondria. Every cell in our body is therefore a kind of mutualism among what were once independent living organisms, now working in some form of long-evolved co-operation. Each mitochondrion still even has some of its own genes, although most of them have since migrated to the nucleus. This kind of co-operation has been so successful that the large organisms that we see in our daily lives are built mostly out of these kinds of cells. Whales, bears, cats and dogs, trees and water lilies, and so forth. They all have this kind of co-operative eukaryotic cell.

Multicellular life is another co-operative venture

For a considerable period of time, that is to say, for several billion years, it seems that life consisted mostly of single cells. The oceans were filled largely with individual prokaryotic and eukaryotic cells. Then, something changed. Cells began to aggregate into a wide variety of sizes and shapes to form larger organisms. This explosion of life was one of the defining characteristics of what we now call the Cambrian Period, that began about 570 million years ago. The quite sudden appearance and diversification of multicellular beings presents a kind of problem. It would seem that for several billion years, single-celled organisms ruled the Earth. So, what triggered the increase in complexity of living organisms? Why did multicellular organisms arise?

There are many possible explanations, and we are now sorting among them. This is how science works. We come up with a list of possible explanations for a phenomenon and then set about collecting evidence to find out which of the explanations are wrong. I frequently had to remind my classes that when you are trying to find an answer to a problem, the first challenge is to make a complete list of the possible explanations. This is just like a murder mystery: you cannot try to solve the murder until you have a full list of suspects. Otherwise, there is a tendency just to pick the first possible candidate and accuse them. This is not a good way to catch murderers or to solve scientific problems. So first, you need a list of the possibilities, then the collection of evidence to discriminate among them. When all the incorrect ones have been disproved, the remaining answer is likely the correct one (at least until someone comes up with an even better explanation and even better evidence). Eventually, consensus is reached, but it often takes many years of work.

Hence, while we know that multicellular life did arise and can date it rather well, the cause(s) of the event has not yet been settled. One reasonable suggestion is that it took several billion years for photosynthetic cells to produce enough oxygen to raise dissolved oxygen levels in the ocean high enough for respiration to evolve. Respiration is far more effective than fermentation as a source of energy, but it requires oxygen. From this perspective, multicellularity is a consequence of aerobic respiration, which itself is dependent upon oxygen, which in turn was dependent upon photosynthesis. All of this took time. Another suggestion is that multicellular organisms arose in shallow water and that it took that much time for there to be sufficient ozone to shield the shallow water from ultraviolet bombardment. The ozone layer requires oxygen in the atmosphere to form, and so the ozone layer too would have been dependent upon photosynthesis. The ozone layer then not only depends upon photosynthesis but upon oxygen leaking from the ocean into the atmosphere, which requires careful consideration of how much oxygen was produced by single-celled plants, and how it cycled early in the Earth's history. A third suggestion is that there may have been a change in oceanic chemistry, near the beginning of the Cambrian Period, which permitted the deposition of calcium carbonate required to make shells and skeletons. These are just three of the more obvious possibilities. The real explanation for the sudden diversification of life remains a puzzle, the sort of situation that challenges scientists to keep working. I usually tell my students that the oxygen hypothesis strikes me as being quite reasonable. We certainly know that photosynthesis evolved long before multicellular life. We can date that event because the production of oxygen changed oceanic chemistry. One marker was the rusting of the Earth's oceans. Biologically-produced oxygen removed iron from the oceans, creating huge deposits of banded iron between 2.2 and 1.8 billion years ago. Most of the world's iron (and hence steel) comes from banded iron formations, which means that metal objects in your life actually have a tangible connection to the earliest photosynthetic cells in the Earth's oceans.

All of the above explanations rely in some way on changes in the nature of the early Earth caused by living organisms. Since the evolution of photosynthesis caused enormous changes in the oceans and the atmosphere, it seems reasonable that the evolution of multicellularity is in some way related to those changes. Yet there are other possible explanations, such as the first appearance of eyes. In *Life Ascending*, Nick Lane describes the evidence that all the many kinds of eyes found in nature can be traced back to a common ancestor. Yes, even creatures as different as humans and lobsters had a common ancestor, as shown not only by our use of serotonin in our brains, but by the similar ways in which we detect light. One of our common ancestors may have been the trilobite, whose fossils are commonly displayed in museums as an example of early deep-sea life. Trilobites appear rather early in the Cambrian explosion of life, but even though this was half a billion years ago, eyes with image forming lenses were already present. If, as it seems, the ability to see originated that long ago, it is just possible that the evolution of eyes and sight may have contributed to the rapid diversification of life at this time. And, as for trilobites, while they are abundant as fossils and while they dominated the oceans for a period of time, they became extinct about 250 million years ago.

Great extinctions

There also have been periods when life suddenly declined. Five such mass extinctions are recorded in the fossil record, and we are now, as mentioned earlier in Chapter 7, seemingly entering another period of mass extinction. This time we know the cause: human overpopulation. And, we have a growing list of the species at risk, the Red List of Threatened Species mentioned in Chapter 1, which currently lists 26,500 species. Some are already gone, of course, like the passenger pigeons that used to nest in my own country, Canada. Others, old friends of mine, like the humble spotted turtle and Blanding's turtle, both of which are residents of wetlands in southern Canada, are now on the list as well. Their habitat, wetlands, is being drained for agriculture, filled to allow expanding urban sprawl, and chopped into isolated fragments by highway construction. Of course, if we know the cause of the current extinction, we know how to prevent it. The most important factor, as described in Chapter 1, is the urgent need to protect the world's last large wild places from further human encroachment. It would be supreme irony if one of the world's successful multicellular life forms, *Homo sapiens*, created so many copies of itself that it destroyed the last of the Earth's wild places and the many kinds of wild creatures that lived there. Yet any chart of human population growth shows that the copying process is continuing relentlessly as our last wild places become raw materials for constructing more human bodies and altering the land to feed and house them.

Getting back on topic, much more could be said about the history of life, particularly the invasion of land and the transitions from amphibians to mammals. How can you write about the history of life without at least mentioning dinosaurs? There. I mentioned them, but you will have to read about the history of multicellular life elsewhere, in a book like *The Earth Through Time*, which sits beside me on my desk this morning. The reason for picking out eukaryotic cells and multicellular life for this chapter is that both of them arose out of natural selection (thank you, Darwin) and yet both clearly involve considerable amounts of co-operation. At the risk of being repetitive, both of these events demonstrate that we should not base our understanding of the natural world on a mere caricature of natural selection.

It is easy to slip into a worldview in which we think natural selection and the history of life involve only competition and predation. We slip into Tennyson's language that nature is "red in tooth and claw." It is so easy that even scientists tend to do so when writing textbooks. Back in 1990, I went through standard university texts that were being used to teach students the principles of ecology. I found that in counting the number of pages on competition, predation, and mutualism, there was a stunning discrepancy. The average textbook had some 30 pages on predation, 50 pages on competition, and less than 10 on mutualism. I noted in one of my articles at the time that eukaryotic cells, multicellular organisms, mycorrhizal fungi, insect pollination, and seed dispersal by animals, are all examples of mutualism. The title of the article was deliberately intended to be provocative: "Is mutualism really irrelevant to ecology?" Apparently, some people think so, because the problem persists. In one of my own books, *Plant Ecology: Origins, Processes, Consequences*, I have carefully devoted an entire chapter of 38 pages to the topic of positive interactions, and so if you are curious about this topic, you can read more about co-operation there.

The world contains co-operation

In conclusion, in order to understand ourselves, we have to understand and appreciate evolution by natural selection. And in order to understand evolution by natural selection, it is necessary to understand and appreciate co-operation. All of our activities as humans are possible only because individual cells gave up something of their individuality and specialized to varying degrees in order to carry out more specialized functions within multicellular beings. In the case of the red blood cells, part of this co-operation involves even giving up nuclei entirely. We exist only because cells can co-operate.

The topic of co-operation actually, and perhaps unexpectedly, loops us back to Chapter 5 and the topic of hierarchies. Why? Because a hierarchy is a kind of co-operative venture. This is one reason why they continually appear in biological and ecological systems. According to Margaret Heffernan (p. 109) "Hierarchies, and the systems of behaviours that they require, proliferate in nature and in man-made organizations. For humans, there is a clear evolutionary advantage in hierarchies: a disciplined group can achieve far more than a tumultuous and chaotic crowd." Teamwork and hierarchy were once likely advantageous for hunting large animals and harvesting crops; now teamwork and hierarchy are necessary for running large airports and maintaining power plants. Each mitochondrion has had to give up a little of its own independence in order to function within a cell, each cell has had to give up a little of its independence to work within an organism, and each organism has had to give up a little of its independence to work within an organization. Just how much can be safely given up and just when we cross the border from co-operation to tyranny, are topics for political scientists and philosophers.

Individuals, cycles, and interbeing

On today's excursion we mostly focussed on our own bodies which turn out themselves to be communities or colonies comprised of many other living beings that are co-operating with one another. They began their co-operative venture long ago in the Earth's early oceans. That imaginary 'self', the me and my body view of the world, is indeed an illusion.

If instead of looking inward and we were to look outward, we would again discover that we are not a single entity. We began looking into the body because that story illustrates how co-operation is a natural part of our being. Yet, for completeness, let us see what happens if we look outward at how we are connected to the world around us.

We are also part of a larger entity and ecosystem, and we are constantly exchanging materials with other living beings, chiefly, as we saw in Chapter 1, the material elements CHNOPS. The carbon, C, that comprises most of our body is largely stored in the atmosphere and in the soil, and we are temporary homes for those carbon atoms. As you read, atoms of carbon in your body are being exhaled in your breath back into the atmosphere. (The atmosphere naturally has carbon dioxide, of course. Levels have changed over geological time and have now reached concentrations higher than 400 ppm. I say now because we have records of atmospheric composition going back a million years, back through multiple ice ages and interglacials, and none of those periods had carbon dioxide levels higher than 300 ppm. Something has changed, but that is another story.) That exhaled carbon from your most recent breath may drift for a while in the atmosphere as part of a carbon dioxide molecule, but then it will be absorbed by a plant. If you live in the city and have a big tree outside your bedroom window, you will eventually have a great deal in common with that tree. Some of it has been built out of your own exhalations. And, should you occasionally pee on the roots of that same tree, it will also have some of your nitrogen and phosphorus to complement the carbon. Hence, what is now that tree was once you. All living organisms exchange matter, and scientists have spent nearly a century carefully tracing how this happens and how fast the exchanges occur.

Instead of putting our attention on what we imagine to be ourselves, and how atoms move through us, let us move our attention to one single atom and follow its individual path. Consider phosphorus, an element that is common in human bones and teeth. A few years ago, each phosphorus atom was in the soil, perhaps inside a fungus, a bacterium, an earthworm, or one of many hugely abundant but obscure soil animals like nematodes and springtails. Then it was absorbed by a plant root, perhaps spending some time above ground in a leaf or seed. Then it was fed to an animal and spent some time flowing around in that animal's blood, perhaps coming to rest in a muscle or bone. Then you ate the meat, and that same atom of phosphorus entered your body for a while, perhaps coming to rest in a bone or a tooth. I know some of you are proudly vegetarian; in that case you may have ingested the phosphorus most recently from a plant but even so, earlier in that same atom's history, it was also likely in a soil animal, and millennia earlier, it might have once been part of a giant ground sloth or a fellow human being. Eventually the phosphorus atom will pass out of our body in urine or excrement, in which case it will find its way back into the soil, or increasingly, into the nearest river or lake. If it is still lodged in a bone or tooth when you die, it will return to the cycle with a certain delay depending upon the fate of your corpse. The Buddha himself was cremated within the Ganges River watershed, and the ashes from similar cremations still contribute a steady flow of recently-human phosphorus into the Indian Ocean. The mortal remains of Charles Darwin, in contrast, were interred in Westminster Abbey, which rather delays his re-entry in the global phosphourus cycle. Getting back to rivers, even in the absence of crematoria, one of the most obvious effects of having humans in watersheds is higher levels of phosphorus in the water. Really. Population density predicts phosphorus content. This is the collective result of thousands and millions of toilets flushing into the river or, worse, street defecations, not to mention the soil erosion from building ever-new expanses of housing. All these phosphorus atoms drift downstream where they are eventually absorbed by algae, some of which might be eaten by zooplankton, and then by fish, and then another person, and so the cycle continues. Increasingly, with more advanced measuring techniques and ever higher human populations, we can even detect caffeine, antidepressants, and human hormones similarly flowing down our rivers toward the ocean.

I would prefer to have my own body eaten by a wild animal, since I know that our large birds and coyotes are desperately hungry in the winter. Even the smaller woodpeckers eat carrion, if they can find it. Alas, exposing my corpse for wild animals would not be legal in my own culture. So, if I should die on this hiking trip (and on the last steep section of the trail, it felt like I might) please leave me a while for the turkey vultures and carrion beetles to enjoy. Not wolves, alas, as they have been mostly exterminated in this part of the world. There have been efforts to reintroduce Red Wolves to the Great Smoky Mountains, but success is not certain, and, in the interim, vultures and beetles are dealing with carcasses.

It is not just phosphorous, of course. The same process is true of every other element in our body. We are all aggregates of atoms, each one having been a part of another plant or animal many times in its life. It is entirely possible that one of the carbon or phosphorus atoms in your body also was once part of Jesus, or a Roman centurion, or Judas. Or the Buddha, or his loyal assistant Ananda, or a peacock that once passed them on the road to Vulture Peak Mountain. The carbon just flows on, in and out of the atmosphere, and in and out of our bodies, driven by photosynthesis and respiration. Thich Nhat Hanh often talks and writes about life as *interbeing*. The process of nutrient cycling shows that this is not just some obscure Buddhist philosophy. It is a statement that is consistent with hard science. You can even calculate the rate at which different parts of the biosphere interchange elements with one another, that is, the rate at which interbeing occurs. Thankfully, I no longer have to mark exams on this topic.

Now a true story, which is a great way to pass time when one has stopped to camp after a long day on the trail. Years ago, I had the good fortune to hear a public lecture by Ian McHarg, author of *Design with Nature*, in Halifax, Nova Scotia. The book was considered a landmark in how to plan human settlements in a way that is minimally harmful to wild nature.

The book is mostly common sense. It says, for example, that we should not build cities where there are important agricultural lands, significant forests, aquifers, or endangered species habitat. The fact that such a simple message was seen as a hugely important message, is, with hindsight, perhaps a statement of just how rapacious humans have been. We even build houses in floodplains and then complain when they are flooded. We build our cities on fine agricultural land and wonder why food becomes expensive. We build houses on steep hillsides overlooking a lake and wonder why the lake fills with phosphorus and becomes green with algae. *Design with Nature* had a huge impact. Sometimes people are ready for some common sense. Then again, sometimes they are not: a local city flooded this spring, and there was whining and hand-wringing in front of the cameras with no one apparently being willing to admit that they had made the foolish choice to buy a house on land that was in a floodplain and had flooded many times before. In my book *Wetland Ecology*, I describe how one of the world's earliest written stories, Gilgamesh, includes vivid depictions of a human city in a floodplain being, well, flooded.

Now for a story within a story, one told by McHarg himself, in his Halifax talk. I have often used it in my class lectures, since the topic of matrix models for nutrient cycling, and actual calculations can be a dry topic. I used this story for entertainment in between the equations and now share it with you. McHarg presented this story as a way to educate our own leaders about the reality of the human condition. That reality is our interbeing, our interconnectedness. Perhaps if our leaders understand interbeing, they will behave with more common sense and perhaps even some compassion. The message of interbeing is important, since, as we have already discussed, most people are so self-centred that they cannot imagine that they are indeed, as just described, already one with everything. That is why these people focus on the 'me' feeling, drive enormous vehicles, steal from one another, kill wild animals, dump trash in the forest, drop bombs on each other, and do all those other annoying things that make samsara the painful experience it is. So, McHarg said, imagine taking a politician you really don't like (I'm sure you can think of a least one if you try hard enough.) and put that person in orbit in a space ship, along with a tank of algae to provide food and oxygen. At first, the politician will see themselves as different from (and superior to) the algae, and maybe, there being no other audience, will even lecture the algae on how much more important humans are than algae. Each day, the politician's waste will go into the tank of algae, and the algae will provide more food and oxygen for the politician. As the days pass, the orbiting politician will turn into algae, and the algae will become the politician. McHarg said that it may take a long time, but when this person realizes that they and the algae are really one and the same being, then, and only then, will it be safe to let them return to Earth. The audience loved the story and so do I. So, this year, when the Iranian Mullahs were shrieking (again) that it is time to exterminate the Jews, I imagined transporting those Mullahs to a space ship where they could cast their imprecations at a tank of humble algae. How long, we might wonder, until they accept the uncomfortable truth that they and the algae are one and the same, like the rest of us? I know, I know, for some people it will take longer than others.

Nutrient cycling does show in a direct and factual way why we really are one with nature. And, if this is not remarkable enough, as was mentioned in the second chapter, all those elements in our body were formed by exploding stars. In the song "Woodstock," Joni Mitchell reminds us "We are stardust, billion-year-old carbon, ... and we've got to get ourselves back to the garden." Those lyrics could be adopted as Buddhist liturgy to be chanted each morning. Another popular description of nutrient cycles can be found in a classic little book, *A Sand County Almanac*. The word 'little' matters in this context, since a paperback version is light enough to slip into a backpack, and tonight I will let our guests share a short essay called Odyssey, which traces an element called simply X (probably phosphorus) as it passes from a limestone ledge, through a flower, a deer, an Indian, a bluestem, a fox, and many more prairie cycles. Aldo Leopold says "An atom at large in the biota is too free to know freedom." Here he sounds a little like Thich Nhat Hanh or like Gary Snyder. Still, most of us persist in thinking of ourselves as individuals. It is a very convincing illusion. It is also just a habit.

Let us get back to the theme of co-operation. We don't want to fall into the trap of childish wishful thinking, that because our cells co-operate, surely then we humans can all live in harmony. A brief study of human history, including some examples shared earlier in this book, shows plenty of conflict. If you think co-operation is that easy, read the history of the Bolshevik revolution or simply try to find a Buddhist community that is not riven by dissent.

There is one more reason that I drew your attention to those two early stages in the origin of life. They illustrate a Buddhist point, that the concept of the self is rather arbitrary. When you stop on the trail and take a mouthful of water from your canteen, just who is it that is doing the drinking? What exactly is the 'self', when each of us is demonstrably constructed from a large group of cells? And what exactly is the self when a significant proportion of our cells do not even have the DNA which we assume defines us as an individual? And what is the self when each of our cells contains mitochondria that originated as free-living prokaryotic organisms? Since we are a combination of so many kinds of cells, with such a long evolutionary history, it is quite remarkable that we could label this community 'me'.

The whole situation is so contradictory. We feel like individuals and in certain ways we are, just like a lobster, an elephant, or a peacock. Simultaneously, we are also colonies of co-operating cells, just one large community. Simultaneously again, we are a small and transient cog in an enormous planetary nutrient cycle. We are atmosphere, ocean, and rocks. Try holding all of these views in your mind at once. They are all true. It might be called a kind of Zen koan. A koan is a statement about the nature of reality, a statement that seems to defy rational analysis. It is said that the study of such koans is one path to enlightenment.

10
Meditation in Action:
Seeing Through the Simulation

> According to the Buddhist tradition, we don't get new wisdom,
> nor does any foreign element come into our state of mind at all.
> Rather, it is a question of waking up and shedding our covers.
> We have these goodies in us already; we only have to uncover them.
> (*Trungpa 1991, p. 5*)

U<small>P TO THIS POINT</small> in the book we have been focused on the human condition and particularly the First Noble Truth, the experience of suffering. We have also given Darwin a fair amount of attention. One reason for doing so is to bring some practical realism to the prospect of a spiritual path. When Chögyam Trungpa taught, I recall that he made it very clear that all we have to start with is who we are. Indeed, he used to explicitly mention 'ape instinct'. In the same way, Suzuki Roshi used to emphasize that we have to start where we are. And where we are now is a long way from the savannas of Africa where our minds were formed. As these two important teachers emphasized, if we are curious about being liberated from illusion, we certainly do not want to start out with wishful thinking about human nature.

We are now going to look more closely at the path itself. Our mountain hike has a landscape ahead, and we use a map to navigate and stay on the Appalachian Trail. The main trail is crossed by many others, some going to more isolated locations and others taking us off the path and into towns. The practice of meditation also has a path of sorts and does require a map and time. It also requires us to pay attention along the way so we don't get lost. The Buddha did not say that reading one book or doing one weekend of meditation would suddenly enlighten you, making it possible to have magic powers and live happily ever after. Rather, the Buddha laid out a path. As we know from hiking, sometimes we get on a wrong path, and then we have to find a way back to the main trail. The Buddha's instructions keep us on the Eightfold Path. Following the translation used in *The Historical Buddha*, these are: Right View, Right Resolve, Right Speech, Right Action, Right Livelihood, Right Effort, Right Mindfulness, and Right Concentration. Right livelihood is a good place to start thinking about the path. When choosing a profession, we should consider what will be beneficial to ourselves and others and what will be harmful. So, making illegal drugs in your backyard, killing bushmeat, or logging old-growth forests would probably not be good career choices. In my opinion, other poor aspirations would include being a rap star or a professional athlete. Such careers may not cause direct harm, but at best they distract people from reality.

We can also think about the consequences of all our actions. Another practical example: the world does not need more huge sports arenas. Promoting them and building them is a harmful activity. It wastes time, energy, money, and materials that could be used for more important purposes. The world also does not need more cats, and the decision to spay your cat and keep it in indoors would qualify as a positive activity. Indeed, one of the greatest threats to wild birds is the level of predation by cats kept by humans: in the United States alone, biologists estimate that free-ranging domestic cats kill between one and three *billion* wild birds each year. Generally speaking, it would also be appropriate to avoid other activities, like parking in handicapped parking spots (unless you really are handicapped), shoplifting, spraying graffiti on walls, or insulting your neighbours. On one hand these are meant to be amusing examples; on the other hand, it really is true that the Eightfold Path calls us to think about our day-to-day actions and how they affect those around us. Do you really need to drive a huge SUV, live in a monstrous house with granite kitchen counter tops, or fly to the Caribbean for your wedding? We contemplate the Eightfold Path and apply it to our lives.

Of course, discernment is necessary. Self-righteousness can become a kind of intoxication. It is a delicate balance, to evaluate and to see clearly, yet not become overly critical. Apparently, some of my hiking companions on this very trip feel that I am judgmental. They could be right. I explained to one of them yesterday that my purpose was to clearly explain the situation, to save wild places, and to help people avoid suffering, not to win a popularity contest for most amiable hiking companion. My concern lies with the fate of wild nature and all other sentient beings. I enjoy human companionship, but if it takes some blunt speech to save whales, rainforests, and wild birds, so be it. I have read Jeremiah, and I know what happened to him.

I offer, as an aside, that perhaps certain companions have taken offense because the truth is troubling: they may own a luxury car that is not even a hybrid, they may have many expensive tattoos, they want to fly to a wedding at a Caribbean resort, they want to sell their family property to a developer rather than donate it to a land trust, they don't appreciate the effort of carrying a pack in the mountains, or they can't be bothered to simply pay attention to the dawn birds because they are too busy complaining about the chilly air. Or maybe all of the foregoing. Taking offense is sometimes a clever way to avoid personal responsibility. In any case, our expedition will be over soon, and somebody has to voice unhappy truths, even if some of our hiking companions find me judgmental. I speak for all the wild beings who have no way to speak for themselves. I am aware that the Buddha himself taught us to avoid 'harsh speech'. But if you claim that any statement that causes someone offense is 'harsh speech', it means that we cannot talk sensibly about important problems. A simple example: I once had a hiking companion take offense when I warned him he was neglecting proper care of his feet. Sure enough, some days later he was bemoaning his blisters.

Part of the Eightfold Path is, of course, meditation. We are offered a practice that allows us to experience the contents of our own minds and to begin to explore our attachments to our illusory self. Instructions on meditation have been passed from teacher to student for 2,500 years. Although basic meditation instructions are now available in books and in various kinds of recordings, we are told that there is still no substitute for personal initiation. I agree. The practice is best passed on from one person to another. It is the same way we learn how to hike and camp and recognize birds and trees: from other people in our community.

However, in order to continue our conversation with Darwin and the Buddha, it will be necessary to provide some written introduction here to practices of meditation. In one way, it is quite outrageous to presume to write a short chapter on the Buddhist path. The field is enormous, and there are many better-qualified teachers. However, one could say it is equally outrageous to write a short chapter about four billion years of evolution and the origin of multicellular bodies. And, yes, there are many people who know far more about Buddhism, such as senior teachers and professors of religious studies. But alas, I don't know of a single one with more than a passing knowledge of Darwin, or an interest in how an understanding of human evolution might inform our understanding of the Buddhist path. I at least have personal experience with both.

There is one further danger in writing about meditation. It is dangerous, because fundamentally, meditation is unlike any other human activity. It undermines habitual thought patterns. It therefore undermines our illusions about reality, the ones that create our habitual relationship with the rest of the world. Since ape ego resists this process, it may be all too hasty to provide its own distorted interpretations of the process. It is indeed quite possible for the ape ego to willfully misinterpret the instructions because it understands that the process is undermining its own existence.

As for the prospect of enlightenment, what could that possibly mean? I have found in my writing that some people, particularly scientists, find the word downright annoying. In the context of this book, we could say that enlightenment could be described as awareness of, and freedom from, the illusions engineered into the psyche by selfish genes. It is different enough from our ordinary experience of reality that it is indeed hard to describe. It has been described this way:

> Of course, when we talk about freedom, we are not talking about overthrowing the head of the state or anything like that: we are talking about freedom from the constriction of our own capabilities. (Trungpa 1991, p. 5)

The three stages of mind training

In the Buddhist tradition overall, we can recognize three cultural traditions of practice: Theravada, Mahayana, and Vajrayana. Each provides a view of the world and certain practices to help us escape habitual patterns of thought and experience. Each draws upon different resources within the mind to reduce suffering and provide liberation. Each also could be a path in itself.

Sometimes these three cultural traditions are presented in a linear sequence. First a focus on self-liberation called Hinayana (the Theravadin tradition, also known as the Lesser Vehicle), creating the ground for a more comprehensive and compassionate practice (Mahayana, the Great Vehicle), with Mahayana then creating the ground for a further stage still (Vajrayana, the Diamond Vehicle). This view of a progression through stages is particularly popular, of course, with those who are practicing Vajrayana, and it is often they who call Hinayana the 'lesser' vehicle. But even the Theravadin practices have developmental stages, as you can read in Bhikkhu Bodhi's guide, *The Noble Eightfold Path*. "Wisdom unfolds by degrees, but even the faintest flashes of insight presuppose as their basis a mind that has been concentrated, cleared of disturbance and distraction." (p. 14).

It is generally true that instructions in Vajrayana (also called Tantric Buddhism) will not be given without students having first completed the necessary foundation practices of Hinayana and Mahayana. For this reason, tantric practices are sometimes treated as secret practices; they are considered to be dangerous to students who have not completed the preliminary training. The secrecy, it should be emphasized, arises at least in part out of compassion for other human beings. Given the tendency of the human psyche to protect itself and enslave its owner, one must be careful not to give it tools that could be turned against its owner and used to reinforce its habitual patterns of control. Tantric practices could do exactly this without thorough preparation beforehand. Trungpa Rinpoche likened tantric practices to grasping bare electrical wires: a skilled electrician might be able to do so in certain circumstances, but to encourage every member of the public to do the same would be to invite tragedy. In the Buddhist tradition there is even a word, *rudrahood*, to describe the consequences that arise when the practices end up reinforcing rather than releasing the bonds of ego. Rudrahood is not only unpleasant for the practitioner, it is a kind of megalomania that is dangerous to the rest of society (not to mention to Buddhist communities themselves). There is an even simpler explanation for what appears to be secrecy: learning has stages, and there is no point in trying to explain how to interpret the subtle autumn plumages of warblers to a student who is just learning to tell the obvious difference between a crow and a blue jay.

Although for a secular reader, this chapter will have a Buddhist tone, recall that 'Buddhism' is simply a word designating a particular body of knowledge about human nature. Hence, it applies to all humans. Thomas Merton, a Catholic monk, was willing to see the commonality in all paths that liberate the mind from egotism. In his Asian journal, written shortly before his accidental death, one reads

> ... in the cultivation of an inner spiritual consciousness there is a perpetual danger of self-deception, narcissism, self-righteous evasion of truth. ... in striving to live entirely by his own individual will, instead of becoming

free, man is enslaved by forces even more exterior and more delusory than his own transient fancies. ... It is in surrendering a false and illusory liberty on the superficial level that man unites himself with the inner ground of reality and freedom in himself which is the will of God, of Krishna, of Providence, of Tao. These concepts do not all exactly coincide, but they have much in common. It is by remaining open to an infinite number of unexpected possibilities which transcend his own imagination and capacity to plan that man really fulfills his own need for freedom. The Gita, like the Gospels, teaches us to live in awareness of an inner truth that exceeds the grasp of our thought and cannot be subject to our own control. (Merton 1975, p.352-353)

A digression on ethics and evolution

Before we enter this discussion of the path, we also need to remind ourselves that the study of human nature and human evolution should not be presumed to provide justification for harmful behaviour. Natural selection and evolution are simply descriptions of biological reality. They are, in Buddhist terms, the source of samsara. They are in no way intended to provide moral guidance. People often miss this point. Just because our ape ancestors once did such and such, it is incorrect to conclude that it is necessary, or even acceptable, for us to continue doing so in modern times. However, all too often, this is what happens. According to some biographers, Darwin himself realized this risk, and it may be one of the reasons he delayed publication for so many years.

Social Darwinism has been one unfortunate consequence of Darwin's teachings. That is to say, conditions such as the terrible Victorian situation of a very few leisured rich exploiting vast numbers of starving poor has sometimes been justified by the dictum of the survival of the fittest. It may be true that the world of biology and selfish genes is indeed a world with survival of the fittest. But this does not mean that we must consciously accept these processes, or build a society that reinforces them. Were we to simply adopt the processes of biological and evolutionary forces to guide our behaviour, we might not drill wells for clean water, use soap, invent antibiotics or carry out surgery either. All of these could be means to avoid the survival of the fittest and the struggle for existence.

There are few, if any, who would suggest that we should stay dirty and not take medicine, just because they alter the rules by which evolution has operated for aeons.

To repeat: the discovery that ape minds have certain habitual tendencies engineered into them by selfish genes does not lead to the conclusion that this is the way humans must behave. We have an inheritance, but it is up to us to decide what to do with that inheritance. The path provided by the Buddha to escape mental conditioning provides the same type of freedom as soap, water, and medicine. Indeed, when stepping onto the Buddhist path, it is traditional to first regard the teacher as a doctor who is prescribing a medicine to assist in dispelling troubling illusions.

Hinayana

The basic teachings of Hinayana are the Eightfold Path and the practice of shamatha-vipashyana meditation. Right Conduct is an essential component of the Eightfold Path. It includes taking an oath not to harm others, including oneself. Whether one can be of benefit to others is left for future consideration; at this point the practice is mostly to cease causing harm. There are guidelines for conduct, all involving the avoidance of habits that harm others. As we have seen, the list of activities that cause harm can be rather long and personal.

Rules are of little use, however, unless one has the correct tool to clarify them and apply them to life experience. This tool is shamatha-vipashyana meditation, sometimes translated as mindfulness-insight meditation. In mindfulness meditation, our objective is to develop a clear and dispassionate perspective on the flow of thoughts and feelings in our own mind stream. Normally we identify with these thoughts and feelings; we call them 'our' thoughts and feelings. The belief that we possess these thoughts and feelings is part of what gives them such power over us. During meditation we constantly relate with the breath, and we discover that the flow of thoughts can continue without us buying into them. (It is not, by the way, that breathing is in some way magical; it simply provides a neutral reference point that we can use to train our mind.) Thoughts flow just as stomachs secrete digestive juices; the thoughts are no more or less important than another physiological process. At first we find that we continually get involved with our own thoughts, but from time to time we wake up and become aware that this has happened. When we wake up, when we notice this gap in thinking, we become aware that we have been asleep. We label the experience 'thinking' and then return to the relaxed perception of the breath. As we practice this (and it literally is a matter of practice, just like training the mind and fingers to play a musical instrument), we discover that the gaps become more frequent and that the awareness of the thoughts and feelings passing through the mind increases.

> We begin to lose the reference point of self-consciousness, and we experience the environment of practice and the world without bringing everything back to the narrow viewpoint of "me". We begin to be interested in "that" rather than purely being interested in "this". The development of perception that is penetrating and precise without reference to oneself is called vipashyana in Sanskrit… (Trungpa 1991, p. 134).

It is like starting the exploration of a darkened house with a flashlight that operates for only a few seconds; as time goes on the flashlight turns on more and more frequently, and as we practice the length of time of illumination increases. Eventually, the period of illumination is long enough to see what the nature of the room truly is. For those who do not practice, our description of the room may appear entirely fanciful. This is a personal process that provides some illumination to see the interior of the mind. We do not have to like what we see. Neither do we have to condemn what we see. It is what it is. We end up, at least in principle, with rather fewer illusions about ourselves and about the world.

The Darwinian perspective upon this is rather straightforward. The small, day-to-day mind can be regarded as a large computer simulation that contains a number of elements. The first element of this simulation is the sense of I-ness, the sense of being an 'I'. The second is the sense of a stage upon which this 'I' appears and acts. And the third is the sense of 'that', the entities with which we must develop strategies for survival. 'That' might be an animal to kill, a mate to seduce, or a king to overpower. Fourth, there is a story line, some internal drama. All this is embellished with various thoughts and emotions. The simulation then runs different scenarios over and over again, offering various possibilities for the way in which the 'I' might interact with the others on the stage. The advantages to having such a simulator are self-evident. The better an organism's simulator, the more likely it is to make the right choice in finding resources, locating and seducing a mate, or rising up the ladder of power within the social milieu.

Humans sometimes enjoy the process of simulation and value it enough to hire others to create various simulations in which they can participate. What, after all, are Shakespeare's collected works but a series of programs examining the possible consequences of love (*Romeo and Juliet*), confusion (*A Midsummer Night Dream*), lust for power (*Macbeth*), and the sharing of resources (*King Lear*)? A great programmer like Shakespeare is valued because of the vividness of the simulations he creates in human minds. Indeed, all of human literature from the earliest heroic poems to the most recent bestseller can be regarded as programs that humans enjoy processing through their simulators.

The problem, which the Buddha called samsara, seems to be that this simulator frequently becomes so powerful that it overpowers us. Then, instead of enjoying our lives, or experiencing reality, we experience only hour after hour of confused thoughts about various kinds of threats and possibilities. In this kind of confused mind, there are only brief instants of escape. A striking sunset, a fine meal, or a lover's glance may wake us up to the here and now—ah! And then the program runs again. The gap in thought may be so small that we don't even notice; the simulator just resets itself and starts a new program. Habitual complaints about the world are also a kind of simulation. These are just stories about what the ego does not like about its surroundings.

When we sit in meditation on a long retreat, we largely remove ourselves from external stimulation and duress. Sitting hour after hour on a cushion with few external stimuli, we find that the simulator runs all the same. We may physically be sitting on a red cushion in a large meditation hall with a timer ringing a gong so we can walk for a few minutes each hour, but in our minds we are still hunting for things, trying to seduce lovers, and seeking status. These illusions continue hour after hour. Sometimes the power of illusion is so strong that only when the gong is rung by the timer do we wake up out of the daydream. This is one purpose of meditation. Sitting in isolation from normal day-to-day action helps highlight the activity of simulation. Otherwise, the day-to-day events of the world, our simulation, and our reactions become mixed together.

Sitting on a cushion for endless hours provides incontrovertible proof that the simulations run almost continually. That is one reason why meditation is difficult. It is actually not all that pleasant to watch the stories our minds tell us about our lives and about the world. True, they are just stories but many also include emotions, and the whole show can be irritating at best and painful at worst. The benefits of Hinayana practice are, however, the ability to see the simulation while it is in progress and to recognise that the simulation obscures our appreciation of reality.

Not all traditions place so much emphasis upon sitting meditation. Indeed, as our hiking trip in the mountains is showing us, walking through the forest can have a similar effect. It depends upon our attitude and preparation. The Japanese Tendai school focuses upon walking meditation: one introduction can be found in the book *Walk Like a Mountain*. Even the Buddha and his entourage spent a lot of time walking from one town to another.

Whether we are sitting or walking, the experience of waking up from the simulation is akin to waking up from a dream and discovering that it is only a dream. It may have been a happy dream, it may have been a sad dream, it may even have been a nightmare; in any case, we suddenly realize it for what it was—an illusion. At this instant, we see the world as fresh and unconditioned, raw and full of possibilities. There is perception, naked, as it is. Period. The purpose of meditation, from the Hinayana perspective, is this: to train the mind to see the world as it is, unencumbered by the illusions superimposed by the simulator. The more we notice that we are awake, the more we wake up. The more we wake up, the more liberated we are from illusion.

As we begin to wake up, we may uncover our own Buddha-like qualities, or, more simply, our own Buddha nature. Note that I have managed to write this far without having to even mention Buddha nature, much less attempt to define it. It is one of those words that has crept into the English language. We may intuitively grasp the intention of the word, and it is possible that the more we try to pin it down with logic and grammar, the more we may confuse ourselves. The Sanskrit term, *tathagatagarbha*, is not terribly helpful, either. 'Oneness' might be a better word; 'interbeing' another possibility. Example: when you are hiking through the forest, are you experiencing oneness with the mountains, the forest, and the calling birds, or are you dwelling on your own petty concerns, such as wet feet or a perceived insult about your cooking? (Yes, maybe you really should have brought your boots into the tent last night, as I advised, since the dew can be heavy in the mountains, particularly when the sky is clear, and yes, maybe your vegan chickpea casserole really is not as appetizing to others as you seem to believe it is.)

Perhaps there is even room for petty concerns within the vast experience of oneness, so long as we do not dwell on our discomforts, and ruminate upon our complaints. After all, Siddhartha Gautama himself did a lot of walking, and one assumes that he too was capable of feeling discomfort from wet feet or the occasional unpleasant meal. In fact, both tradition and *The Historical Buddha* suggest that he died from food poisoning. (Which could take our minds back to ruminating on that chickpea casserole.) Staying on the topic of Buddha nature, here is what Suzuki Roshi has to say on the matter in *Zen Mind, Beginner's Mind*: "When you understand a frog through and through, you attain enlightenment; you are Buddha."

It is easy to imagine that meditation is a life-long practice that is slow to produce effects. Sometimes it is even taught this way, perhaps to reduce expectations. We are probably not likely to discover our Buddha nature or enlightenment suddenly, although Zen masters assure us that the possibility is always there. Suzuki Roshi's classic book is, after all, titled *Zen Mind, Beginner's Mind*. However, if we suddenly decide that we have become enlightened and start telling our friends so, it is probably time for some further guidance from an experienced teacher.

Instead of looking ahead to enlightenment, whatever that might be, Hinayana practices also have short-term benefits. Sameet Kumar has written about his own experience with Hinayana practices in a book titled *The Mindful Path Through Worry and Rumination*. He has a particularly challenging set of clients: people with cancer, many of whom have a terminal condition. Their 'lifetime' of practice may be quite short. Yet he describes how short daily sessions of mindfulness improve the quality of their lives. He explains how many of us have a hostile inner voice that encourages us to ruminate on the unhappy aspects of our existence. He reports that some of his patients discover that in spite of their difficult circumstances, "present conditions offer a different beauty, joy, and happiness, which we can enjoy in the immediate present if we have the mental presence to do so." Kumar also has more to say about the physiological changes that occur as we rewire our brains. Hence, while we can take on mindfulness as a lifelong practice with the aspiration for full enlightenment, mindfulness itself apparently can have beneficial short-term effects too.

Often people will say that their lives are simply too busy: finding even half an hour a day for mindfulness practices is just too hard. Here is the strange thing. Once one begins to practice mindfulness on a regular basis, one notices just how much time in our lives we really waste. Consider some obvious examples that we see around us: time spent on hobbies, time spent watching television, time spent on sports events, time spent on the internet, time spent playing video games, time spent on the telephone, time spent daydreaming, and so on. It would seem that we often fritter our life away.

It is much the same in the secular world. Some of my biology students used to make the excuse that they could not learn the names of the wild plants and animals in their region or visit important natural areas because they were just too busy. Often these students were engaged in one or more of the preceding activities. Really, an honest assessment of our lives would suggest that we waste enormous amounts of time. We can, instead, make the decision to reallocate some of our time. Moreover, basic mindfulness practices will often help us find even more time within our apparently busy lives.

Mahayana

As we develop some capacity in meditation, we can engage the Mahayana path and practices. Here, the emphasis shifts somewhat, from thoughts to emotions and from self to other. In one way, the Mahayana experience arises somewhat naturally out of Hinayana. When we practice shamatha-vipashyana meditation, we start with noticing and labelling our thoughts. They are just thinking. But there are a lot of them, and they seem to arise endlessly. Nonetheless, if we stick with our practice, some sense of peacefulness is possible, at least from time to time. We may experience periods without so many thoughts. And we can see that the many thoughts we have about our life are just thoughts, not reality itself. The Mahayana encourages us to pay closer attention to emotions and their associations with habitual thoughts. We find that large clusters of thoughts are often associated with certain emotions. These emotions themselves have a raw energy of their own. Pema Chödrön uses the analogy of weather. There can be wild weather where she lives in Cape Breton, with cliffs rising up out of the ocean. We find that we also have inner weather, with emotional fronts passing through. We can experience them as weather, and maybe even become adept at riding them like wild ravens or eagles. Consider resentment. One gathers from online commentaries that many people are troubled by storm fronts of resentment. When resentment arrives, it is often covered in thoughts that provide that emotion with justification. Or, rather than a storm front we can use the analogy of waves. Waves can be covered by the froth and garbage from past experience. We have the opportunity to see these emotions for what they are, waves of energy that carry a set of images and memories. Like waves, they eventually subside, or crash on the beach, only to be followed by others.

Overall, we find that at a more subtle level, two kinds of unconscious processes are operating in our lives. We seek to pull pleasurable experiences toward us and we seek to escape painful experiences. This is a tendency we share with even the most simple sea creatures, crawling toward food and away from threat. There is an important psychological consequence for humans: our world is thus habitually divided into two states. There are things we want and things we don't want. Thus we are (unconsciously) being dragged through life, alternatively attracted and repelled by phenomena. Most other living beings have similar experiences. We are not that different from lobsters or starfish, let alone from our fellow mammals or primates. Another human example of how this plays out moment by moment. "I really would like a glass of red wine with dinner, perhaps a Merlot. I really hope that annoying neighbour does not turn up his stereo when I sit down to eat." Hope and fear.

The Mahayana path also challenges us to extend our vows. With our minds now awakening to our habitual patterns, we are inspired to take a vow to not only avoid harming others but to help others. This inspiration arises out of seeing the degree to which our own behaviours are engineered by habitual emotions. So our view of the world becomes larger, and there is greater attention to our surroundings and to other living beings.

There are new styles of meditation that accompany this view. One, called *tonglen* in Tibetan, is somewhat less well-known, but again instructions have been published, and so there is no harm in sharing them with readers here. Of course, the intent is to provide some insight into the path. If you intend to do the practice, you should obtain formal instructions from a qualified practitioner. We begin a tonglen practice session by recalling the degree of openness that exists when the mind is not spinning habitual stories. We then deliberately arouse painful stories and experiences. The general instruction is to recall some event in another's life that was very painful, or some image that arouses sadness. As these negative feelings arise, we now do something utterly astonishing; instead of pushing them away, we breathe them in. We reverse exactly the psychological process that our ego has been using for so many years. Some students are actually afraid to do so, for fear of poisoning themselves with painful thoughts, and these fears illustrate the degree to which our unconscious mind secretly fears the poison of the outer world. The process is then reversed; positive feelings of joy and happiness are aroused. We might do so by recalling the joy of being free of negative thoughts, or recall some pleasurable event or image. Once we can feel the texture of raw joy we do something equally outrageous; instead of holding on to it, we breathe it out and give it to the rest of the world. This procedure is then repeated. We train to allow pain to ride on the in-breath, and release from pain on the out-breath. We breathe in human negativity and suffering; we breathe out joy and happiness.

Over time, this practice not only clarifies how pain arises in our mind, but it builds the capacity for compassion. The ability to experience compassion may be regarded as part of our enlightened inheritance, but it is an ability that may have atrophied under the pressures to acquire and succeed. So we begin to regularly practice compassion. Since compassion is an inherent part of ape instinct, we are simply starting where we are by using something we already have.

When we first begin the practice of tonglen, we can be troubled by just how difficult it really is. It becomes clear that the ego is constantly afraid of feeling suffering and continually craves the feeling of pleasure. It is so determined to continue doing this that it can actively resist the process of tonglen. As we gently persist, we discover that the practice seems to loosen our habitual patterns of behaviour. When someone comes to our desk and insults us, we are no longer quite so conditioned to fight with them; we can simply feel the pain that lies behind their anger.

Although such practices may be unfamiliar, perhaps this brief introduction illustrates how it might be possible to strengthen natural tendencies that already arise in our ape minds. Mahayana practices may indeed help expand the compassionate side of our human nature and help us escape from our tendency to harm others.

At the same time, it is apparently not that easy to escape ape instincts. In tonglen, the instruction is often to begin with compassion for our closest relatives and then expand our view outward to those we know less well and those we dislike. Starting with our closest relatives is starting with the worldview of our selfish genes, although I have never heard that said anywhere else. I can also say that I have never had the instruction to extend our mind out even further, to embrace all the wild creatures in the world. As I have been taught, tonglen mostly stops with fellow human beings. Why? I see no reason why this should be the case. It seems that we could teach tonglen by taking in the suffering of the world's whales, rhinos, and all manner of wild creatures. It just does not seem to be happening, at least in the teaching situations I have experienced.

Selfish genes also remain strong, in my experience. Here I have some real data. At meditation centres people put out lists of individuals who are suffering and who might benefit from tonglen. There are other online sources where people solicit tonglen practices for those who are suffering. Over the years, I have seen many hundreds of entries on such lists, and I can assure you that nearly all of them are for close relatives! I may be the only person in our community who sees tonglen requests and immediately sees evidence of selfish genes speaking. I have so far resisted the temptation to write northern white rhino, California condor, or leatherback sea turtle on such lists mostly because I know that I am indulging my irritation and creating confusion for people who apparently just don't get it. The human names on those lists, however, show bluntly that most Buddhists with whom I associate may be doing their utmost with compassion for other human beings, or at least their close relatives, but at best, their compassion mostly does not extend beyond humans. Since the objective of this practice is to dissolve the boundaries between self and other, and to enhance our capacity to feel compassion for other, there is every reason why whales and rhinos, if not butterflies and mushrooms, should be included. We might also gently remind people that the world is bigger than their own immediate family.

Back to Darwin

Now for some input from Darwin. We understand that in order to maximize survival and reproduction, organisms have to grasp 'good things' like food and mates and to avoid 'bad things' such as hunger and loneliness. The problem seems to be that this grasping and avoidance process causes us pain. Moreover, it is somewhat out of control because the program runs even when it is unnecessary. Hence, we feel the compulsion to grasp or avoid when, actually, nothing needs to be done at all. Sometimes it is so difficult to just sit quietly and watch the sun rise over the forest. As remarkable as it seems, much of the frustration and dissatisfaction in our lives arises out of a struggle with reality, or more precisely, our struggle with our imperfect simulation of reality. Perhaps in our distant past, when we were marine invertebrates or more recently when we were apes on a savanna, this behaviour was necessary. Now it is eating us alive. It has stolen our freedom. Instead of experiencing true freedom, instead of appreciating our lives, we are driven by habitual patterns of behaviour.

> Wanting to grasp the ungraspable,
> you exhaust yourself in vain,
> As soon as you open and relax this grasping,
> space is there—open, inviting and comfortable.
> (Nyoshul Khenpo 1993, p. 4)

And now, Vajrayana

Vajrayana practices arose, it is said, when an early King asked the Buddha for practices that could be applied in his everyday life since he was unable to assume monastic vows or go on long retreats. Vajrayana is considered to provide the most rapid means for achieving enlightenment. Of course, the people who tell us that are usually Vajrayana practitioners! So, we might retain some scepticism when we hear such claims.

From one perspective, the Vajrayana path simply adds to the arsenal of tools available to subdue the ego. From another perspective, the Vajrayana approach is distinctive in that it employs the very experiences of the simulator to achieve enlightenment. An analogy might be the many martial arts where one uses the momentum of one's opponent against them; here the simulator itself can become part of the practice of liberation. Only a few elements of Vajrayana will be mentioned here and only to illustrate the degree to which these too mesh with Darwinian reality.

On the Vajrayana path, the practices of mediation are expanded by the application of visualisation practices, usually visualization of deities who are said to assist in liberation. As we have already seen, such deities have a psychic existence as natural kinds of energy within our own mind. As powerful archetypes, they provide a source of psychological energy beyond the reach of the conscious mind. From the Darwinian perspective, we know that archetypes, as defined by Jung, are a natural part of our psychological structure, and in Vajrayana the power of these very archetypes is, at least in principle, used in waking up. From this perspective, we might appreciate tantric deities as embodiments of the energies that we have inherited from our ancestors. In this sense they are indeed sacred.

A controversial aspect of this approach is that it also proposes to take advantage of some of the features of the world that may normally be employed by the ego to keep us asleep. Behaviours that normally keep us securely wrapped in a cocoon of habitual thoughts might also be used as tools to wake up. Alcoholic beverages, for example, are commonly used to dull pain and deaden feelings; from a tantric perspective, alcohol might be used to intoxicate the simulator. Similarly, a sudden surge of anger or fear that might once have felt overwhelming might instead stop us in our tracks and create awareness on the spot. In short, from the Vajrayana perspective, everything that occurs can either be part of the simulation, or experienced as a part of awakened mind. The task of practice is to harness the power of the phenomenal world and the texture of psychological illusion in order to wake us up, on the spot.

While visualization sounds quite exotic, it too is part of our human nature. Our inner monologue and its emotions are often accompanied by visuals, that is, images. Some images from the past can be very strong indeed. Here is an example from music. In the song "Diamonds and Rust," Joan Baez describes how an unexpected phone call from "a booth in the Midwest" takes her back in time, "Now you're smiling out the window of that crummy hotel over Washington Square." With a few lines, she whisks us from a scene in the Midwest to a scene in New York's Greenwich Village and from her present phone call to her past experience. And she evokes all the pain she still carries from those days in New York with her old lover. It is a lovely song about human suffering altogether and how we carry it forward in our lives.

The same process of visualization can be used to cultivate chosen states of mind. Most of us can think of someone who has inspired us to live our lives in a better way, and some may already have a photograph of that person in their home. Perhaps it is someone like Gandhi, or Martin Luther King, but often it is someone more personal. Near my office hangs a simple framed black-and-white photograph of Suzuki Roshi. Over my dining room table is an artist's rendition of St. Francis of Assisi. When we see such images, we may be inspired. We actually would like to have some of the qualities the person seems to possess. We can work with this desire as a style of meditation. If we sit somewhere quiet, we can deliberately call up an image of this person. Usually it is best to begin with a period of shamatha-vispashyana meditation, to help us settle into the practice. We then use our mental image of the person to remind us of the qualities they inspire in us. That is, we are invoking the energy they represent. We visualize these qualities flowing toward us and into our own bodies. And, at the end of the session, we can allow the image to dissolve into our own heart. The dissolving part is important: it is only a visualization.

You can see then that this style of meditation is a little more complicated than sitting on the cushion and noticing the rise and fall of the breath. We are actually engaged in cultivating certain qualities. We are merging our normal sense of 'self' with something that we normally regard as 'other'. There are further instructions for this process, which you can be taught on another occasion. We may, for example, visualize the person in front of us or above our head, or we may visualize our own body as being that person. We still end the session by allowing the image to dissolve into our heart.

This is a powerful and possibly transforming process. And, of course, it need not be limited to people. Those paintings you see in meditation centres (or at museums or art galleries) depicting strange looking deities like White Tara or Medicine Buddha can be used in the same way. We could aspire to be like White Tara or Medicine Buddha. These are not real external beings, at least in my opinion, but they represent certain kinds of enlightened qualities. White Tara, for example, has multiple eyes, including one on the bottom of each foot, which symbolize that her awareness extends in all directions. In order to use such a deity in visualization, you have to be introduced to the deity image by someone who already knows the deity and who can explain what the symbolism means. Otherwise, it won't work. I mean, there would be no point in visualizing White Tara thinking that you will end up with eyes on the bottom of your feet and wondering how you would walk on the Appalachian Trail in that condition. The deities symbolize kinds of energy. Sometimes geometric designs, such as mandalas, can be used in the same way. And, there is also a bit of magic here, if you wish to use that word. The introduction to the deity is provided in person by a qualified teacher who was themself introduced to the deity in a similar process, and so this process takes you back perhaps millennia, perhaps to the times of the Buddha himself. Authorities disagree on the origins of certain deities, and while I know where certain birds like Bicknell's thrush spend the winter, I don't know where these deities have come from, except to specify in space that it was mostly Asia and in time that it was mostly the past. But there are living lineages of human beings who will continue to introduce you to deities, should you wish to be. There are often preparatory practices, like a required period of shamatha-vispashyana and certain rituals. This process is known as transmission.

Real humans also can be represented as deities. In my home here in the forest there is also a representation of Chögyam Trungpa as a deity. He is seated on a blue horse and surrounded by many colourful phenomena. Some are deities. Another part of the image shows a Himalayan building in which he spent a critical period in retreat. Overall, the image is expansive, elaborate, colourful, and a reminder that there are many kinds of portraiture. Compared to the stately portraits you see of past kings or presidents, this is a very different way of representing a human being.

If this all seems rather exotic, allow me to point out that the familiar Catholic Church has similar practices where one visualizes the Madonna or Jesus. And all those images of saints! Take Saint Frances of Assisi, who is often used to represent a person who loves wild nature. Paintings of him can evoke a powerful sense of our capacity to care for wild animals, and some of the paintings that I have seen come very close to those of Buddhist deities. In *Francis and the Wolf* by John August Swanson, the representation over my own dining room table, St. Francis appears with a bird perched near his right shoulder and a wolf resting at his feet, surrounded by other images depicting animals, landscapes, and stories from his life. Some days it feels like he really is present in the room.

The Eastern Orthodox Church still has beautiful images of spiritual qualities in their tradition. These are known as icons. In the Tibetan Buddhist tradition, we call such images *thangkas*. As we know from history, icons are controversial. Some Christian sects despise icons because they believe that a person can end up worshipping a particular painting or statue instead of the deity it represents. Sadly, Europe is greatly depleted in such religious imagery because, in general, when Protestant armies captured Catholic churches, they tended to destroy all the religious imagery. So the European cathedrals we visit as tourists are likely only skeletal remnants of their medieval beauty. Sometimes you still see the empty niches that once held religious statues. And we probably cannot even imagine the loss of icons that resulted from the recurring Islamic invasions of Hindu and Buddhist cultures in India. There is a reason why so many of our Buddhist teachings survived in the mountains of Tibet rather than in India. The process continues: Muslims recently used artillery to destroy the sixth-century statues of Buddha in Afghanistan. So, few of us have experienced a culture in which we are surrounded by icons. Museums and art galleries have only fragments of what the world once looked like, and even there the artefacts are presented out of context in denatured settings, where they often speak mostly about our cultural sense of how a museum or art gallery should be organized. Most of our homes have art that is intended to bring momentary pleasure rather than spiritual guidance. Mass-produced art can be bought even in furniture stores. In this sense, Vajrayana practices may therefore seem quite removed from our normal relationship with images. All the same, visualization is a process that any of us can use to cultivate positive or inspiring qualities. We can also use images to empower our homes and offices.

As you may have noticed in museum collections of Buddhist art, there are also angry deities, although they tend not to be displayed as often. They too have a purpose. For example, you could visualize a person you despise and engage with that energy. I would advise against this unless you have had a good deal of previous experience. Some trails are simply dangerous for beginners, particularly if you already have a heavy pack on your back. You can see, however, that such a practice would be effective at exposing our more uncomfortable emotions. If you visualize a person who has hurt you in the past, it is remarkable how quickly painful emotions arise. Even our heart can begin beating faster, and we may find ourselves clenching our fists. We are seeing an intense part of ape instinct, the desire to harm those who have harmed us. Here is another scary part about this particular style of visualization. All of us are already doing so all the time! Think about the amount of time we spend ruminating about people who make us angry: an ex-wife, perhaps, a boss who abused and fired us, a friend who betrayed us. We are already visualizing and doing so in a harmful way. We are actively cultivating our negative emotions. So, while it might be insightful to visualize harmful people, perhaps, for short periods, mostly I would stay away from it unless you are working closely with an experienced teacher. Speaking from my own experience, I find that there are enough irritating people in my own reality (like neighbours who drive all-terrain vehicles through my forest) that I do not need to bring up imaginary people to annoy me. Positive visualization practices, on the other hand, would seem to have a great deal of potential to better our lives and our conduct. While it would be inspiring if St. Francis were still around, visualizing him is all we have. An icon or thangka can help. The good news is that if we keep working at our practice, we may begin to embody these very people in our own lives. In that sense, we are bringing their positive qualities back into existence.

An assessment

The Buddhist path in general makes demands upon us, and the Vajrayana path in particular has risks. It may not be suited to everyone. Chögyam Trungpa, for example, described a problem he encountered with western students: painted images of strange deities sometimes seem to attract people who are mentally disturbed. He describes how when he first began teaching in Great Britain, some people arrived who needed some form of mental health care, not meditation. In *Born In Tibet*, Rinpoche confessed that he really did not know what to do with such people. Prominent psychologists like Carl Jung also had their reservations about visualizing Asian deities. We now have some further decades of evidence showing how this path has so far influenced westerners, and overall, it seems possible that these practices can actually cause harm to susceptible people, teachers included.

So, before we consider benefits, let us honestly consider some costs. One cost, or set of costs, is obvious. In ecology and economics, we call them 'opportunity costs'. Roughly, the meaning is that if you are putting time into visualizing deities, particularly for many hours daily, it is time taken from other activities, such as keeping a job or maintaining human relationships or working with a local land trust. Some of the chaos in Buddhist communities is legendary, although it is not always easy to decide if the practices themselves created the chaos, or if they simply attracted people who were already having trouble navigating reality. Another opportunity cost is a divorce from biological reality. To put it bluntly, if people are spending hours and years of their lives visualizing deities, they do not have these hours to take a canoe trip down a northern river, or hike the Appalachian Trail, or learn to identify the birds in their forest. Further, when they then try to teach others, they do not have this rich body of experience to draw upon and share with students. The senior teacher who told me he would not learn his local birds because it might interfere with his ability to commune with the spirits was perhaps trying to tell me about his own opportunity costs.

Another opportunity cost may be the tendency toward introversion. In our lives we have a complex mixture of outer and inner experience: for example, the (outer) encounter with a fox in a meadow is accompanied by the (inner) pleasure that the experience creates. It seems possible that some people become so engaged with illusory deities that they begin to prefer them to actual human experience of the outer world. Wild nature has its own agenda: a wild fox will appear of its own volition and then inconveniently walk away out of the meadow following his or her own instincts. In contrast, a visualized deity appears more or less at personal command. Here there are considerable risks: what happens if people become so involved with their deities because it provides an excuse for disengaging from the natural world? And what if contact with the natural world is necessary for our sanity? And what of other living beings in that 'outer' reality? The survival of big animals in particular and wild places in general requires people to be engaged with them, and if compassionate people are ignoring the natural world because they prefer their inner reality, it does not seem like a positive outcome to me. Might this also explain the absence of teachings addressing our natural world, even in Buddhist role models like Shantideva?

Another cost of this process is the unfortunate tendency to collect deities. It is not just Buddhists who have this problem by the way. Stamp

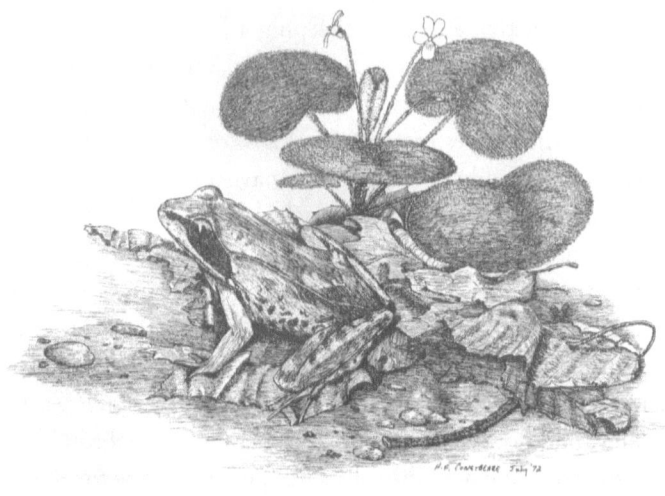

"When you understand a frog through and through, you attain enlightenment; you are Buddha."

Suzuki Roshi,
Zen Mind, Beginner's Mind

collectors are another group. Bird watchers too. I have friends with life lists of birds which they are only too happy to talk about. They will travel a long distance just to add a bird to their life list. Conversation can be quite irritating in a group of these such listers. The same with Buddhists. You can and do hear people sitting and discussing their deities and their empowerments. There is a strong sense of one-upmanship, if someone can announce he has a special transmission for a deity that no one else in the room possesses. Fortunately, having been with bird watchers, I am prepared to feel irritated by deity listers. I know how it feels when someone tells me that they flew to Puerto Rico just to see an endemic bird that I will never see. Yes, I did go up into the mountains once to see a Puerto Rican tody, and all I saw was a banaquit. Banaquits are common on many islands. But although I did not see that rare tody, the banaquit was a lovely bird, and I can visualize it now. The point is that deities can easily become just another kind of bird watching, and easily become an ego trip. Instead of flying to an island for a rare bird, you fly to Tibet for a rare deity. There is even a kind of class structure that evolves. You may be at a meditation centre, and only those students with a certain transmission are allowed to enter a room and practice together. One understands how this is necessary. Nevertheless, it creates status hierarchies in Buddhist communities.

Perhaps there is reason that Trungpa Rinpoche told his students "You should go camping." Perhaps he should also have told them to join the Audubon Society and learn their birds. It is possible to look around one's Buddhist community and meet people who seem to have fallen into themselves and lost touch with the natural world entirely. The Buddhist path could become just another excuse for ignoring biological reality and just another kind of egotism. Indeed, I have had the personal experience of hiking in the Colorado mountains with younger Buddhists who were so busy chattering with one another about their deities that they did not appreciate the Lodgepole Pines or the Rocky Mountain Jays. In this respect, my human companions were no different from people in airport waiting lounges who want to talk about sports teams or Jesus. The Buddhist path had become, it seems, just a different story line blocking out their experience of reality.

Since all of these Buddhist practices involve working with the contents of our own mind, they do have inherent risks. Overall, the world is full of troubled people, and it is possible for them to become more troubled by some practices. Even normal people may find that encountering certain emotions, memories, or instincts can be uncomfortable. And we know from history that some people are easily persuaded that the world has witches, demons, and angels. You have probably seen people walking down the street engaged in conversations with them. As noted above, Chögyam Trungpa himself wrote about his problems with such people coming to study Buddhist meditation. Gautama Buddha had rules for people who would be admitted to his community and, even so, had difficulties with certain students once they were admitted. Overall, it is probably unwise to dive into complicated practices for looking into the human interior without a good advisor, some common sense, and some scepticism. But this is not just true of meditation. The same is true in the external natural world. One needs some sensible precautions for hiking trips, canoe journeys, and even short, guided excursions like whale watching. People get mauled by bears, or get lost, or fall overboard and drown, or have heart attacks from modest levels of exertion. These unfortunate events are not valid reasons for refusing to hike, canoe, or whale watch. They are reasons to be sensible and cautious while doing so and to be certain that you learn how to put up a tent or paddle a canoe from someone who is well-qualified. The same advice applies to extended meditation retreats or visualization practices. Whether you go outward into wild nature, or inward into wild nature, both can be unforgiving to those lacking respect, skill, and common sense.

Finally, we could look at outcomes. Practice should have outcomes. If all these practices work as well as we are told, if we all have Buddha nature and people have been practising for decades, then we should by now have a good cohort of enlightened people in North America. This would seem to me to be a good thing, if only that it means one no longer needs to fly to Tibet to seek advice. As a scientist, I am trained to make predictions, and then test them against reality. I understood Gautama Buddha and his many modern representatives teaching in North America to say that enlightenment was both desirable and possible. Yet, overall, I do not see large numbers of enlightened people appearing and walking out east across the plains from the Rocky Mountains to restore peace and wilderness to the world. (I say walking east because Suzuki Roshi taught on the west coast and because the Rocky Mountains have the Great Stupa of Dharmakaya, an enormous monument to Chögyam Trungpa, built from a special concrete mix intended to last a thousand years.) Of course, in saying this, it is also possible that I simply lack the degree of enlightenment necessary to see them. A great Indian Buddhist mystic called Naropa spent many years looking for his teacher, who, it is said, in one case simply appeared to him as a dead dog. Or so the story goes. Fortunately for him, he persevered and did finally meet Tilopa in person on the banks of the Ganges, and that was the beginning of the lineage that produced some of my own teachers. Getting back to the present, it sometimes seems that people with many years of practice have become detached from western society and western norms. As you may have gathered, I am not impressed by teachers who have been born in North America and yet have limited knowledge of wild nature, or human history, or western society altogether. Nor am I alone in perceiving that the Buddhist path is tending at times toward recreating Asian theocracy and that the benefits of this trend accrue mostly to those at the top of the hierarchy. With the passing of time, it is also evident that some of our most senior teachers have displayed ape instincts such as the desire for power, the craving for alcohol, and rampant indulgence in sexuality. What does such behaviour have to say about the potential for accurate transmission of the Buddha's teachings in the west? In his book *How the Swans Came to the Lake*, Rick Fields has compared our Buddhist teachers to wild birds. Perhaps such birds cannot thrive outside their native habitats of Asia.

So, on a personal note, I can say that although I have practiced for many years, and although I know of many people who have practiced with much more diligence than me, I have not yet seen countless Buddhas (or even a St. Francis, for that matter) arriving in the western sky on flaming clouds to save our whales, rainforests, and coral reefs. Or, if those expectations are too high, to save at least our grizzly bears and Blanding's turtles. The people who inspire me, like Guy Bradley and Tommy Douglas, remain few and far between and are mostly not Buddhists. Perhaps the lesson for me is to lower my expectations. There are countless other Buddhist teachers whom I have not heard or met in person. Or alternatively, perhaps the lesson is that people who call themselves Buddhists should take their own vows seriously. Buddhist practices should not be used to escape from reality and avoid the urgent need to protect wild nature. But then, as noted, some people on this hike have already suggested that I am too judgemental. It might be so. Still it does not invalidate my observations.

Escaping samsara

The Buddhist path does help us work skillfully with our inspiration to make the best of our lives. It is apparent that one of our most important challenges is to realize that what we take for reality is really a cloud of simulations rather than reality itself. With daily mindfulness sessions and other kinds of practice, we can take realistic steps to reduce the impact of such simulations in our lives. This means coming closer to an understanding our true nature. We are told that this true nature transcends our ape ego. It includes the possibility of finding our own Buddha nature, a state of mind beyond aggression and ambition. Here is what another acknowledged tantric master, Nyoshul Khenpo Rinpoche, has to say on this matter. The vocabulary and style of exposition may seem unfamiliar. Note how he describes the ability of awareness to provide liberation from the simulator which, in this translation, generates the 'conceptuality' within which we live. He offers the possibility that we could escape this conceptuality and realize Rigpa, "innate vajra-like Buddha mind."

> Those who recognize their true nature are enlightened; those who ignore or overlook it are deluded. There is no other way to enlightenment than by recognising Buddha nature, and achieving stability in that; which implies authentically identifying it within one's own stream of being/mindstream, and training in that incisive recognition, through simply sustaining it continually, without alteration or fabrication. (Nyoshul Khenpo, p. 35)

> The wisdom-mind of all the Buddhas, innate wakefulness, is inherent to our very nature; yet it is temporarily obscured by conceptuality. Innate vajra-like Buddha mind, Rigpa, is unveiled the moment dualistic mind dissolves and nondual awareness nakedly dawns.... This is the authentic Buddha, the Buddha within. There is no Buddha apart from one's own heart-mind... (p. 40).

11
Enlightened Society:
The Evolutionary Imperative

PERHAPS THE MOST ENDEARING QUALITY of human beings is our desire to make things better and our belief that this is possible. This may have been one reason a small group of people, probably only a few hundred, left Africa about 100,000 years ago and set out across the Middle East. Since then, humans have spread around the world, even reaching Australia and North and South America. Along the way we have built cities, irrigation canals, bridges, pyramids, universities, monasteries, temples, sailing ships, clocks, steam engines, canals, railways, highways, airplanes, nuclear reactors, and rockets. The incessant inspiration to create might be the defining character of the human species. Without it, we might never have changed from wandering tribal primates on the plains of Africa. With it, we have been driven to explore the entire world, build an orbiting space station, and land robots on the surface of Mars. So let us look, finally, at this particular desire for improving our circumstances.

One word we might use to describe this type of desire is 'longing'. This is one of the emotional states that we are told brings people to the Buddhist path in particular and to spiritual paths altogether. We could describe it as the conviction that our lives could be better than they are. This longing extends outward to our society, in the view that our communities could be better. One of the joys of reading small town newspapers is the stream of letters to the editor about how the community could be improved. A cynic might observe that too many small towns think improvement goes no further than building a hospital and a skating arena for the local hockey team. A cynic might also suggest that this longing for something better might indeed just be another manifestation of craving. But here let us consider the proposition that longing also encourages us to take an enlightened attitude toward our world.

But first, a caution. Today's hike will include some magical terrain, including the mythological kingdoms of Camelot and Shambhala. And Blake's mystic vision of Jerusalem. This could all seem rather fanciful, even annoying. So, let me be clear that we are actually talking about something that is quite practical and real: our innate desire to take an enlightened attitude to our world. And to our household too.

In the same way, the Appalachian Trail and the Eightfold Path benefit from such an attitude. Example: if you have ever had the misfortune to hike with a complainer, you will know how just one negative person can drag down everyone's trail experience. It really is difficult to stay on the trail and appreciate the mountains, the trees, the birds while someone else is whining about their feet or the rain or, more often than not, their current partner or their ex. Yes, my friends, on top of all the necessary camping gear and the food, some people will actually carry somebody they dislike for the entire journey. Then they wonder why no one ever wants to hike with them again. More desirable companions take an enlightened attitude. Here is a positive example: the sort of person who, in spite of the morning drizzle, will point out some flowering rosebay rhododendrons in the mist or who will call our attention to the song of a hooded warbler. Another example: the sort of person who will carry the extra weight of a bottle of wine for an entire week, simply so we can all share a cheerful glass together over dinner during our last night on the trail. These are the sort of people who make good hiking companions and good citizens too. So, enlightened attitude actually is a practical topic. To attain such a realization, everyone should go camping.

Longing

In the very first chapter of this book, on the first day of our hike, we discussed craving and how it creates dukkha. Now, with some experience of human nature, let us consider there is, as well, an aspect of desire that is more subtle, more nuanced, more gentle. We will call it longing. It refers to the aspiration that humans possess to find their place in the cosmos, to have a useful role in society, to better themselves, and to help others. This longing to be fully human in a peaceful society is possibly another innate quality of human beings. It may be one quality that distinguishes between civilized, as opposed to barbarian, existence. We can engage with this quality by taking an enlightened attitude toward our own life and toward those around us.

Longing, combined with inspiration, compassion, creativity, and effort, are positive qualities that have produced many of the advances of modern civilization. These enlightened qualities in human beings may even give our life a certain sacred quality. This does not mean that we need to create imaginary deities. It means we need to take a positive attitude and apply our best qualities to living our daily life. We can have an entirely non-theistic basis for this attitude.

A spiritual path does not require us to ignore our rational knowledge of the world, including our humble origins on the African plains. Some people are afraid of secular knowledge, and even deny Darwin's biological discoveries in particular and modern science in general. Yet is it not a kind of sacred view to acknowledge that our bodies are built of stars that exploded billions of years ago, that our arteries carry ancient oceans through our tissues, and that our inspiration to create is the product of millions of co-operating eukaryotic cells? Is it not sacred to see that the politics within our own human communities are really not that different from interactions among our marine ancestors that lived a billion years ago? Is it not sacred to see that we are literally, not just metaphorically, one with everything? (This recalls a Buddhist joke, one of the few I know: the Buddha walks into a pizzeria and says "Make me one with everything.")

Even enlightened qualities may be driven by basic evolutionary imperatives underlain by selfish genes. Let us conduct a simple mind experiment. Imagine twenty small ape tribes. Endow ten with a longing for something better, and let ten be completely satisfied with things as they are. This longing should be general and undefined for a good reason: it is not possible to predict exactly what obstacles these tribes will need to overcome. Will they need to find new hunting grounds? Will they need to develop some new tool for removing marrow from a carcass? Will they need to develop the lever or the wheel? Will they need to discover and propagate the seeds of a Mediterranean grass? Will they need to domesticate grazing mammals? One cannot say in advance. But this much is clear: the tribes that possess longing will likely surmount obstacles in a way that the other tribes will not. In the long run, there is every reason to believe that the tribes endowed with longing will have left more young than those without. Over time, then, longing will triumph over acceptance. It is part of human nature.

The dark side

Longing has its dark side: if apes long for something better, then they also find themselves dissatisfied with their current conditions. A common way we experience this dissatisfaction is a sort of primordial restlessness. When we sit in meditation on a cushion, we directly encounter this behaviour. Who could imagine how difficult it is to simply sit still? We itch, our knees hurt, we want to get up and go somewhere. And, our mind is just as undisciplined as our body: who could imagine how difficult it is to rest our minds on our breath?

As we have seen in other chapters, and as the Buddha himself taught, life is unsatisfactory. The experience of dukkha could also include restlessness. There is the risk that our ape minds will constantly flicker away from the present, away from nowness, focusing instead upon possibilities that tantalize us but hang before us, just out of reach. We might call this the state of primordial confusion that we see on our streets today.

There is another possible dark side to our longing for something better. It is the naïve belief that we can do it by suddenly restructuring our society. The French had their Revolution, and it was followed by The Terror. The Chinese had their Great Leap Forward, and it was followed by famine and chaos. Lenin promised relief from abuse by the Czar and bequeathed us Stalin's murders and the prison camps of the Gulag Archipelago. In the early twenty-first century, we now have people telling us that if we divide everyone into tribes based upon imagined gender and self-identified race, we will have a brave new society. The promises are always for a better world, but experience shows us that changes can also make the situation worse. Perhaps far worse. The mere longing for a better world does not guarantee easy solutions or happy consequences. That is why some knowledge of history and ape instinct might be useful when we plan ways to improve our societies.

A further dark side of longing appears in our daily experiences: wishful thinking. If only we had money without having to work. If only we could live in a cozy little cottage with a pleasant garden, neatly-mown lawn, and white picket fence. This is what my mother wanted, to judge from a cushion top she embroidered in her childhood. What she got was the Second World War. We could update her embroidery by adding the essential elements of a modern suburb: a neatly herbicided lawn, dwarf junipers, a riding lawn mower, membership in the local golf club, and two large vehicles in the driveway. If only we could have these things, we know in our hearts that we would be at peace.

There seems to be this natural and understandable tendency to want a cozy little nest in which to hang out. We hope that we will find it by hiding out in a dream home in a cozy reclining chair. It is remarkable how often that term "dream home" is showing up in the media; it really seems to matter to a lot of people. In my own countryside, I repeatedly see people moving here, only to bulldoze the forest aside and build an urban house complete with asphalt driveway and lawn with lawn ornaments including perhaps a plastic deer. They have left the city only to recreate it, destroying natural forest and wild beings in the progress. We could rebel against such behaviour and desire something different, perhaps a log cabin with solar panels where we weave macramé hangers for spider plants while drinking herbal tea. Oddly, unexpectedly, whatever the images we have of peace, when we struggle to create them, it causes us pain, and if we do manage to create them, they often turn out to be prisons. At best we have created an illusory world that puts us to sleep and lulls us into lethargy. At worst, the very struggle to get rid of the messy, dirty, unwanted parts of life becomes part of our attitudinal problem. This desire to make a cozy nest seems to have its dark side: it is actually a private little bubble in which we try to shut out the world and seal ourselves off from the larger view. Another example: people moving to the country and installing huge outdoor lights that then obscure the stars.

The desire for coziness actually seems to be in opposition to our longing for a better world. One of the oldest books of western civilization, *The Odyssey*, addresses this very dilemma. *The Odyssey* begins at the end of the Trojan War. The city has been sacked, and the Greeks are returning with their booty. Odysseus is living with the sea nymph, Calypso, who has promised him immortality if he stays in her cave. Immortality with a sea nymph in a cave seems to be quite a cozy situation. But instead, Odysseus decides to sail home and reclaim his kingdom and his queen. He gathers his crew and sets sail. He soon encounters the lotos-eaters. The lotos-eaters have no memories and no aspirations, and when one of his crew eats lotos, the sailor forgets his home, his family, and his country; he longs for the endless empty dreams of nothingness. The ancient poets whom Homer recorded seem to suggest that peace was not to be found in lethargy or a cozy cocoon, yet here we are millennia later still seeking peace in the cave of the sea nymph and the land of the lotos. Perhaps Homer understood our evolutionary heritage better than many of us do today.

Blake's vision of Jerusalem

Some two hundred years ago the British romantic poet (and Christian mystic) Blake wrote what could arguably be called the greatest poem in the English language. It has been put to music and sung as a Christian hymn on countless Sundays in countless churches around the world. In this poem, Blake expresses the universal desire to live in an enlightened society. His compassionate warning about 'the dark Satanic Mills' in the first stanza still rings true more than two hundred years later. The last four lines are:

> I shall not cease from Mental Fight,
> Nor shall my Sword sleep in my hand,
> Till we have built Jerusalem
> In England's green and pleasant land.

If the desire for peace, for inner and outer harmony, is a nearly universal longing, why is Blake's Jerusalem so elusive? How can this be? Here is one part of the problem: while human longing for Jerusalem may be universal, no one has given us a trail map showing how to get there. Blake and so many earnest prophets among us sometimes make me think of the parent who gives his child a beautiful new sports car, hands them the keys, and yells "Drive it! Drive it! Drive it properly!" This sort of gift and this sort of advice can easily produce misery and further confusion. We need some better instructions.

Let us therefore return to our hiking companions, Darwin and the Buddha. Both men were both careful observers of human nature. Darwin's observations on our human condition and its origin in natural selection troubled him. What are humans to make of a world where the fittest survive, where we are all in a race to see who is the least impermanent, where our minds lie to keep us in the race, where our emotions continually distort our thinking, where we cannot remember the lessons of our own lives, let alone the lessons of history? From this point of view there is no escape from the evolutionary game in which we are short-lived actors. The good news is that some of our positive qualities, including compassion and co-operation, likely arise out of the same process.

The Buddha did not understand genetics, inheritance, the fossil record, or natural selection. But he did have some tools for dissecting the ape mind and for extricating ourselves from the game. One of the most important tools is the process of meditation. It provides a tool for seeing the evolutionary game for what it is. Meditation is thus a way to experience our own ape minds at work. We experience, moment to moment, how we are creating our own prison, how we are continually obstructing our own search for peace.

Returning to the analogy of the car, we may then treat the Eightfold Path, the practice of meditation, and the many aspects of Hinayana, Mahayana, and Vajrayana teachings as a kind of owner's manual. We start with the keys and are told how to open the door and start the engine. But if we need them, there are other details too, such as how to open the trunk, check the oil, or repair the brakes. What's more, the relationship with a teacher means that there is a personal guide to sit in the seat beside us while we learn to drive. The choice is ours, however. No one can make us follow the Eightfold Path.

Blake's image of Jerusalem illustrates our longing for something better. Just how is it that, with so much money, power and resources, we have more dark Satanic Mills than Jerusalems? And more anthropogenic wasteland than wilderness? Why have the forests and wetlands of the Middle East become desert? How is it that our aspirations are continually frustrated? How is it that peace is transformed into civil war? Some of the blame lies with poets like Blake who told us of the possibilities and then forgot to give us a road map.

Well, to be fair, Blake gave us four lines of instruction:

> Bring me my Bows of burning gold:
> Bring me my Arrows of desire:
> Bring me my spear: O clouds unfold!
> Bring me my chariot of fire!

But what do arrows, spears, and desires have to do with a spiritual path?

Jerusalem, Shambhala, and Camelot

We might turn to the warrior societies of Camelot or Shambhala for one possible path. The example of King Arthur's Camelot has inspired many westerners. Similarly, an enlightened society called Shambhala inspires some Asian societies. Chögyam Trungpa had a particular interest in the legend of Shambhala. He was working on a book on Shambhala when the Chinese invaded Tibet; the manuscript was lost when he fled over the mountains into India. He was transformed from being an abbot and spiritual head of a great monastery to a refugee. At first, he taught the basics of Tibetan Buddhism and Buddhist meditation, but after moving to North America, he eventually began teaching in a different style, presenting a path of meditation firmly rooted in a secular context. He called this view The Sacred Path of Warriorship and wrote a book with this title.

This path was intended to join our aspiration for inner peace with the aspiration for enlightened society. The path begins with the universal human aspiration for a better world. Using meditation as the vehicle, one explores the illusions of ego. One also discovers other human qualities, such as insight, fearlessness, compassion, and elegance. The grip of our illusions is lessened. We see that we are carrying a story line that we impose on the world around us. As this process unfolds, our view of ourselves and our world begins to change. From this larger view, the path to enlightenment is open to everyone, not just Buddhists. We can all develop and cultivate an enlightened attitude.

From this perspective, we can respect our own cultures. Wearing colourful robes, sitting on thrones, and chanting in foreign languages are not a necessary part of the path of meditation. Indeed, they may be outright affectations and distractions. They may even bring unwanted consequences, such as deference to tyranny. Alas, this secular path is currently not available, having been re-Tibetanized over several decades. An example: at one time the teacher wore a business suit and sat on a simple chair. Now the teacher often wears brocade and sits on a throne. An old feudal hierarchy has re-emerged.

The good news is this. Each of us could still take the attitude that the path of meditation can be undertaken with a secular view of the world, with an appreciation for our own circumstances. We can still appreciate the details of our lives, extend ourselves to other people, and experience the natural world as a whole. We do not need the label of 'Buddhist' to make this happen. Nor do we need a leader on a throne. The raw material is still our own murky ape mind. This mind has a fertile mixture that includes aspiration for peace, anger at obstacles, and fear of failure. It is not easy to sit and watch this mind at play. But where else can we start? A path that demands that humans be perfect before starting is no path at all.

It is like our hiking trip. You prepare as well as you can, you find some good companions, and then you start out on the trail. Along the way, you may find some of Blake's burning arrows, and bows of gold, and some instructions on how to use them effectively. At very least, you can be guaranteed that we will encounter mountain views and forest wildlife.

You start on the trail and see what happens. If you get on a side track, you re-trace your steps and continue. You need to keep track of your own map and compass, too, just in case you are separated from your companions. It can be a disaster if you appoint someone as trip leader who is not up to the task. This brings us to the difficult topic of leadership in human societies altogether.

Hierarchies make bad decisions: King Priam and the march of folly

Darwin has reminded us how hierarchies are widespread in nature, so let us explore a little more what their consequences might be for the human condition. We need to understand hierarchies clearly, because people at the top of the hierarchy are making a lot of decisions that affect the rest of us. For every Dalai Lama, there seem to be many more tyrants. Consider just one contemporary example from a theocracy: the Mullahs of Iran are openly stating their desire to destroy the neighbouring State of Israel and drive the Jews into the sea. Crowds are seen in the streets chanting "Death to Israel" and "Death to America." There is one thing we can learn from history, such as the examples in Chapters 5 and 6 show: when people say they are going to exterminate you and take your land and your possessions, it is reasonable to take their threat seriously. Precisely because it has happened so many times before, it can happen again. So, here is one obvious lesson: people at the top of hierarchies are particularly dangerous when their ideology praises the subjugation of their neighbours.

There is an aphorism that also has something to say about hierarchies: power corrupts and absolute power corrupts absolutely. The statement is attributed to Lord Acton, who was an English historian and politician. If we look at enough examples from history, it does indeed seem to be mostly true. That, of course, is the whole point of an aphorism: it is a short statement about the nature of reality. We may look for occasional exceptions and even find them. Perhaps we think of the Dalai Lama, or Gandhi, or the Buddha himself. But the difficulty in finding such examples points out the essential truthfulness of the aphorism. So here is an important lesson for us all: be very, very careful about hierarchies, and even more careful about whom you put at the top.

This observation matters because it is people at the top who are making the decisions for our own future. Whether it is the decision to clear more rainforest in the Amazon, or to drill for oil in the Gulf of Mexico, or to launch missiles at a neighbouring country, or to allow large numbers of migrants into our democracies, it is people at the top of the hierarchy making these decisions.

Here is another general pattern, but one without its own aphorism. People at the top often make notoriously bad decisions. Barbara Tuchman has written an entire book on this topic called *The March of Folly*. She picks examples throughout history where powerful leaders had the opportunity to make the right choice and yet still picked the wrong one. The decision by King Priam to take the Trojan Horse into the City of Troy ranks right up there with great bad choices in history or at least in legend. It offers modern people a kind of archetypical bad decision when there were clear alternatives that were less dangerous. We know how that bad decision turned out. That horse contained Greek soldiers, who escaped at night, opened the gates to other Greeks, and sacked Troy so thoroughly that it took modern historians many years to find its presumed ruins. Indeed, it is still not clear which parts of the story are history, legend, or myth. Some nuances to the story are even more interesting. King Priam was warned on multiple occasions that admitting the horse to Troy would be a disastrous decision. It is sometimes said that it was necessary to dismantle the gates of the city, the very gates that had protected the city through years of warfare, in order to admit the horse. It is said that when the horse was being moved, witnesses claimed to have heard strange noises coming from inside. Some of the king's own advisors simply wanted the horse to be burned. Another insightful person named Laocoön, who tried to warn the king, was attacked and crushed by a sea serpent; for good measure the same serpent attacked his sons as well. (There is a famous Greek sculpture, popularly known as Laocoön and his sons, now on display in the Vatican. It seems to say, in silent marble, that it is unwise to tell people a truth they do not wish to hear.) Another prophetess, Cassandra, also gave a warning and also was ignored. In spite of all the reasons for caution, so the story goes, the Trojans ignored common sense, brought the horse into the city, and left it unguarded while they celebrated. The next day Troy was plundered and the citizens murdered and enslaved. So it was a bad decision to bring the horse into the city. The memory of this decision is still in our language in the expression to "beware of Greeks bearing gifts."

One might think that the story of Troy is an instance of singular stupidity, until you read Tuchman's other examples, up to and including the war in Vietnam. An example for Anglo-American readers: consider how King George mishandled his colonies in the New World. With just a little more common sense governance, the American Revolution need not have occurred at all. In this case, world history was entirely changed by the stupidity of one king. And, as Barbara Tuchman notes, he was constantly getting a stream of good advice from people who knew far more about events in the Thirteen Colonies.

The March of Folly is one of those books that you can't put down and can't forget. And, we humans are now on our own collective march of folly, destroying the life support system of Earth. We have already seen that there are two root causes, one of which is overpopulation. But instead of focussing upon zero population growth, our current leaders are allowing countries where overpopulation has already destroyed the local environment to export their unwanted offspring to other countries. The world cannot support the continual expansion of the human population but today's King Priams are opening the gates to unsustainable migration of human populations as if it were some sort of solution. Natural environment destruction in their home regions has already had disastrous effects for the wild species that once shared those landscapes. That is one important reason why the world's list of endangered species keeps growing. Allowing high rates of population growth to continue and then simply exporting huge numbers of people to less damaged areas will only expand the zone of environmental degradation further. We saw the underlying causes in Chapter 1, which has the only equation in this entire book. You can see the evidence all around us in growing cities and urban sprawl. Wetlands and forests vanish beneath concrete and asphalt. It gets worse: in 2019 my own provincial government is trying to reduce existing protection for endangered species, using the argument that protecting endangered species (and the wild places in which they live) will delay rapid construction of more housing for humans moving to our province. Mass migration, camouflaged by idiot compassion, will destroy our wild nature in just the same way that the Greeks destroyed Troy. It is simple cause and effect.

Why do people at the top make such bad decisions? The first reason possibly is the simplest. It is that the people at the top of the hierarchy are just a random draw from the pool of humanity all together. So, sometimes the person at the top of the hierarchy will be reasonably intelligent, other times they will not be so. It might just be the familiar bell curve in human IQ. If it is just the bell curve at work, there is still a considerable problem, because hierarchies magnify the effects of bad decisions. If one member of regular society makes a bad decision in their life, we are unlikely to remember. If that same person makes a bad decision for their city

(King Priam, Troy) or their country (King George, England) they become legends. In this case, the principal lesson would seem to be that we need to ensure that we have talent at the top. If you read the history of various kings of Europe, there are many examples of talented men having wastrels for sons, which is one of the problems with hereditary monarchies. It should not necessarily be the case: a king should be able to find the very best talent to educate his children, and the very best ministers to guide them. All we can say from looking at our history is that many kings have failed at this task.

There is another possible explanation. Perhaps the talents that get people to the top of the hierarchy are different from the talents that they need to run an organization or a country effectively. What combination of talents might get you to the top? If you are a lobster, big front claws help. If you are an ape, strong front arms and sharp teeth. And if you are a human? That is the question. Given our choice, we would probably like people at the top of hierarchies to be compassionate and wise. It is possible that neither of these traits will actually get you to the top.

There is a third possibility. Perhaps it is not the random draw hypothesis. Perhaps it is not just that the wrong kind of people succeed or, to put it more technically, the traits that allow one to climb the hierarchy are not the traits needed to rule a country effectively. It might be that the very process of getting to the top changes people in some way. People learn from experience. If ruthless acts have been reinforced by success, rulers learn to be even more ruthless. Genghis Khan built a huge empire on ruthlessness and left a lot of children too.

But if you look at the behaviour of some of our current leaders, religious and political leaders both, they frequently seem genuinely confused and out of touch with reality. In some strange way they just don't seem to understand the severity of the problems we face, and the solutions they offer are so obviously ineffective. So, here is a worrying fourth possibility. It is possible that being at the top changes brain physiology. Jordan Peterson points out that animals at the top of hierarchies have brains flooded with serotonin. The fact that they are on top really does make them feel good, as we explored in Chapter 5. And here is the point: they may feel so good that they do not see looming dangers in the same way others do (such as the danger lurking in a wooden horse built by an army they have fought for years). We know that all human beings are trapped within a cocoon of their own illusions, and it is just possible that leaders at the top of human hierarchies are experiencing a different illusion from the rest of their citizens.

All of which suggests that we should be very careful about how we select our leaders, put strict limits on their powers, and replace them regularly.

This is not to say that we should be cynical or defeatist. It seems to be part of our human condition to long for visionary and effective leaders. The reason for contemplating the stories about King Priam in particular and in *The March of Folly* in general, is to remind us not to be guileless. We can seek inspiration without falling into wishful thinking.

Governance

Let us look back at the Buddha himself. *The Historical Buddha*, by H.W. Schumann, to which I have referred repeatedly, provides a good overview for those interested in "The Times, Life and Teachings of the Founder of Buddhism," and it draws upon sources in the Pali cannon. We can say with some confidence that the Buddha himself led by consent. But, of course, he had handpicked his followers, and he had a reputation that they respected. Furthermore, his monks had removed themselves somewhat from surrounding societies and hence, somewhat removed themselves from the dynastic intrigues of Indian society as well as from many of the disruptive influences of money, sex, and alcohol. We might think back to those times and imagine there was harmony. Actually, there were internal problems within the sangha even in the Buddha's time.

Consider the more mature era of his life. In 493 BCE, the Buddha turned seventy; Schumann describes this part of his life as "A decade of crisis" (p. 232-247). According to Schumann, his influence over the monks, "especially the younger ones," was waning. And, yes, even the Buddha himself faced battles for dominance in his community. A monk called Devadatta wanted to climb the hierarchy. He also schemed with a local prince to assist him in seizing power. On one occasion, Devadatta told the Buddha openly that he was old and worn out and demanded that the order of monks be handed over to his control. The Buddha refused and even arranged for one chapter of the community to pass a vote of non-confidence in Devadatta and publicly proclaim the decision in that city. There were also apparently outright attempts on the Buddha's life, although the historical accuracy of these is difficult to judge. On another occasion, Devadatta attempted to split the sangha and become leader of at least one of the resulting fragments.

Meanwhile, in the surrounding Indian society, local leaders fought one another for territory and dominance, and one victor even starved his predecessor to death in a dungeon. Professional armies were emerging which made conflict more efficient but it also led to higher levels of taxation to support those armies.

Gautama Buddha refused to name a successor and justified his decision to his disciple Ananda, saying that the teachings themselves were his legacy and that his students should "live as islands unto yourselves, being your own refuge." But as Schumann says, "The Buddha's funeral pyre was scarcely cool before the monks began asking themselves: What now?" A council was held but that of course required someone to choose the participants. And so, the Buddhist community began to wrestle with politics.

Several thousand years later, debates continue. Some now aspire to be led by god-kings, such as those who once ruled Tibet in the name of the Buddha. Indeed, some of our contemporary Buddhist liturgy refers to the 'omniscient Karmapa.' Regarding Tibet, there have been debates over the degree to which their religious leaders exploited the masses at the lower end of the hierarchy. I will not review these arguments here, except to note that it would be consistent with what we know of hierarchies throughout history: wealth and power accumulate at the top at the expense of those at the bottom.

There is another question, however, which must be asked. I see in my Buddhist community somewhat of a child-like faith in the wisdom of these divine leaders. They are expected to advise on every aspect of the human condition, from marriage counselling, to medical advice, to political organizations, and even to predict the future through divination. It really is necessary, therefore, to ask a painful question. If such leaders were models and were 'omniscient,' why were they unable to appreciate the immediate threat to Tibet posed by the Chinese Communists and the People's Liberation Army? Why were they unable to organize and plan effectively to resist that army? In Chögyam Trungpa's personal description of his escape from Tibet, one is struck by the confusion. Trungpa says his gurus had not given him clear instructions and that "It was beyond them to imagine that the Chinese could take complete control of Tibet and that they would destroy all the monasteries and change the whole Tibetan way of life." (p. 145) What then is the meaning of that word omniscience? Is there any substantial difference between the situation faced by Tibet in 1950 and that facing Poland and France in 1939—a well-armed and dangerous neighbour? We accept that those pre-war leaders of Europe were not omniscient and hence, their countries were invaded and pillaged. Yet some of our Buddhist liturgy and some of our community still state an aspiration for rule by omniscient leaders.

The question remains: what form of governance might guide our communities and societies while we seek enlightenment? A few principles seem clear. The Buddha himself did not name a successor, although he did leave a body of rules for monastic followers. We can say that he trusted his students enough to let them work it out for themselves. He did create a monastic community, with its own rules and regulations, but it is self-evident that we cannot all be monks and nuns, even if we wanted to. Moreover, many of us who are inspired by the Buddha do not wish to follow the example of monasticism. Indeed, one could argue that the world needs engaged citizens, rather than people who emulate the Buddha by walking away from their families and their civic responsibilities to become mendicants. If we walk away, who is going to run the power plants, organize our highways, protect our wild places, and plan our society overall? How are Buddhists in particular and humans in general going to combine their spiritual path and social structure?

It would seem that part of the answer may lie in our own personal conduct. We may long for a perfect society, but the starting point is our own lives. Before we set out to change the world and design new social systems, (and occupy or rebel against everything?) we could begin by effectively ruling our own households. That would seem to be the purpose of the Eightfold Path: how to organize our own lives. In addition, the Buddha had other rules for his followers. From *The Historical Buddha*, these were

1. Refrain from taking life;
2. Refrain from taking what is not given, that is, theft;
3. Refrain from sexual misconduct;
4. Refrain from telling lies;
5. Refrain from intoxicating drinks.

These five rules were interpreted differently for lay people and monastics. Monks were required to be celibate. The topic of what is appropriate behaviour between men and women would seem to be simple, but the continued popularity of advice columns and magazine headlines suggests otherwise. Apparently, words like respect and responsibility are insufficient as guidelines for sexual interaction. Worse, many journalists apparently still do not understand the fundamental evolutionary differences in perspective between males and females owing to their basic differences in biology and that inconvenient Y chromosome. As for intoxicating drinks, they are widely available, and during work on this manuscript, cannabis too became legal in Canada. What would the Buddha say? I really enjoy a glass (or two) of wine with my dinner. Indeed, I consider taking the time to set the table, prepare a nice meal, and enjoy it, along with good manners and good wine, are all a small but necessary part of having an enlightened society. My English mother taught me this and so did Chögyam Trungpa! There are only five rules on the above list and yet it is clear that as simple as they may seem, they remain difficult to apply in this modern era. And, these rules tell us mostly how to behave, but do not deal with larger issues of social organization. Still, the point seems to be that if we are going to have discussions about how to govern ourselves, we would be well advised to start by getting our own apartment or home in shape as a part of this process.

Bhikkhu Bodhi has produced a collection of excerpts from the Pali Canon called "The Buddha's Teachings on Social and Communal Harmony" if you wish to know more about the Buddha's teachings on social organization. However, it really is mostly summarized in those five instructions given above. That is, the Buddha was mostly concerned about how people related to one another as individuals. He does not discuss the relative merits of monarchies, republics, and democracies, for example, nor gives advice on the need for or timing of elections. He does give a short section of instructions for kings, mostly reminding them to care for their citizens. This is obviously good advice, but hardly original. Also there is nothing in his advice to kings about protecting large areas of wilderness or about keeping human populations in balance with wild nature or even about the importance of retaining forests in watersheds to maintain the flow of clean water to human cities. Thus we can generally say that when we are looking for advice on human governance and environmental protection, there is little point in looking back into the past.

Returning to the topic of longing, while we may long for a perfect society, the Buddha left us an open question. Darwin, or more precisely, modern biology, reminds us that there is always a lobster king or silver-backed male ape waiting for the right opportunity. We know from recent history that if we are not careful, we can easily empower a new Stalin or Mao. This takes us, oddly enough, to the American constitution. One can read that constitution as having one central concern: how to organize society so that one tyrant cannot easily seize power. Hence, powers are divided. And rulers can serve only for limited periods of time. The founders of the United States were revolutionaries, but they also had unhappy experiences with monarchial hierarchy. Thus, they bequeathed to Americans a framework that protected against tyranny. Perhaps that is the best we can expect in a political community—the freedom to organize our own households and follow our own spiritual path with a minimum of state intrusion. This leaves each of us with a further responsibility to become involved with the day-to-day running of our communities, however challenging that may be. It might seem rather a poor substitute for a new Jerusalem, or Shambhala, or Utopia, and a divinely-ordained monarch, but perhaps it is the best governance that the Buddha and Darwin would be able to agree upon.

At very least, we can say that when we are sitting in meditation, we are no longer making a nuisance of ourselves. It is certainly preferable to shouting at the neighbours, spraying graffiti on buildings, or making homemade pipe bombs or suicide vests. In this sense, sitting in meditation is already a superior activity. And, according to the Buddha and according to more contemporary Buddhist teachers, the experience of meditation offers a good deal more: a path, perhaps even an open highway, to find some measure of personal peace in ourselves and to cultivate some level of decency within our society.

Let us close by returning to our own human nature. This means we need to make friends not only with our ape ancestors but even with those ocean-dwelling ancestors that preceded them and still live inside us in various ways. This process of making friends with ourselves also entails making friends with all of wild nature because we are one and the same. Our mind is part ape, part lobster, part wild forest, and therefore part of all other living beings. Darwin has helped us uncover the rules that govern these living beings. And, when we sit still and follow our breath, we can directly experience how Buddha nature is part of human nature and how human nature is part of wild nature. We can experience within ourselves the longing to serve the world as a whole. The longing to serve the world allows us to take an enlightened attitude to our life.

This has been a long hike. Now it is over. The paved parking lot signifies that it is time to return to normal life. Our packsacks can be unloaded and safely stored. A mountain hike is different from a book in this way: on the hike, you have no choice but to finish. The only way out is the trail itself. I thank Charles Darwin and Siddhartha Gautama for making time in their busy (after)lives to join us on the expedition, frankly sharing their views, and not complaining about the rain. I trust all of you enjoyed the mountains, the sky, the forest, and the conversations. And, yes, since you

When we sit in meditation, we become aware of our own thoughts. Through this process, it becomes possible to experience the world for what it is, rather than living within a simulation. Hiking can have the same effect: our inner dialogue can be interrupted by a vivid sunset, or by the expansive view across a mountain valley. With time, we can train ourselves to wake up and appreciate even more subtle experiences, such as a bigleaf magnolia, or the unexpected call of a hooded warbler. A single day in the forest therefore offers each of us countless opportunities to wake up to reality.

ask, I did get to see a hooded warbler and a bigleaf magnolia, both vivid reminders of why we need wild places like the Great Smoky Mountains.

In writing this book about Darwin and the Buddha, human nature, Buddha nature, and wild nature, I have aspired to help students of meditation in particular and human beings in general appreciate their human heritage overall. Students may then be better able to find their own way forward to enlightenment. Perhaps along the way they will feel less trapped by their craving or more cautious when they feel drawn to join a hierarchy. Perhaps, too, they will be more effective in bringing about desirable changes in the world and protecting all the sentient beings that live in our wild landscapes.

Perhaps some readers will even achieve enlightenment. Or perhaps you are skeptical and think that achieving enlightenment is far too ambitious for any one human being. In that case, we might choose a goal that is less demanding. We could aspire to simply appreciate our own existence and to be less of a nuisance to our neighbours, particularly all those non-human neighbours living in wild places. That would surely be a good start.

Acknowledgments

First I should thank the many people who taught me biology and who encouraged and supported my determination to work with wild nature. Back when I was still in high school, Bruce McBride corresponded with me about biology with particular advice on making accurate observations. At York University, Duncan Cameron and Michael Boyer not only taught me ecology but also encouraged my activities outside the classroom. Dan Strickland and Dan Brunton shared their extensive knowledge of Algonquin Park, while Ron Pittaway taught me (something) about birds and Tony Reznicek educated me about plant identification. My friends still cheerfully tell me about how little I apparently learned from any of them. In graduate school, both Barrie Goldsmith and Ian McLaren encouraged my extra-curricular work on the natural environment of Nova Scotia, while the Dalhousie University chaplain, Don MacDougall, took time from his hectic schedule to help me polish my backpacking skills and protect natural areas in the province. Back around 1977, Don and I had an astonishing meeting with the Minister of Environment about a karst landscape near South Maitland: forty years later it is still not protected, but at least mapped as pending (The Five Mile River Wilderness Area). That Minister of Environment gave us a stern lecture about how sometimes "bunny rabbits" get hurt on highways; fortunately, I can no longer remember his name. In that same era Don and I, with a few companions, also scouted out a potential extension of the Appalachian Trail into Nova Scotia through Pictou County, and there is now indeed an International Appalachian Trail through that region (www.iatnovascotia.ca). Brian Gifford and Anne Linton (now Anne Greene) assisted me and the Ecology Action Center in an unsuccessful campaign to stop the Wreck Cove Hydroelectric Project in 1975. Sadly, the most southern part of Cape Breton Highlands National Park is no longer a vast peatland with nesting Greater Yellowlegs but merely a hydro-electric reservoir; the good news is that since then, the areas north of the national park are now protected within the Polletts Cove-Aspy Fault Wilderness Area, more than 27,000 hectares.

On the Buddhist side, I wish to thank Joan Kerik for encouraging me to explore Buddhist meditation, and Henry Chapin and Palden McLennan, who spent many hours in the 1980s sharing their knowledge of basic principles and practices. Sonam Rinpoche cheerfully invited me into his home in Toronto to learn about "enlightened attitude." Trungpa Rinpoche, on the other hand, taught indirectly and from a distance. There is an old saying that the best teacher is the one who lives three valleys away. Hence, I have no personal experience of the chaotic party life in Boulder that apparently surrounded him. Nor can I comment on some of the wilder stories still percolating through the internet about that community. Through the main period he was teaching, I really *was* working on ecological research in isolated locations, exploring wild areas by canoe, teaching on campus, or working on environmental issues from Wreck Cove to Great Lakes wetlands. I will say that I often wish that his attendants had served him less alcohol and had instead taken him hiking in the Rocky Mountains. Alas, they did not, he died young, and he left a troubled legacy. Indeed, it sometimes appears that many of his American students, at least those living in his valley, used the opportunity for personal indulgence of every kind. He actually did tell his students there that they should go camping but, in my experience, they have mostly ignored this advice. Hence this book.

My wife, Cathy, was supportive of this writing project from the beginning to the end, starting with our own visit to Cape Breton long before there was either an abbey or an official wilderness area. On that trip we memorably found the elusive curly grass fern, but did not see a Bicknell's thrush. Cathy also has participated in many Buddhist events and retreats along the way, sharing the adventure too. And, yes, we have indeed hiked the Appalachian Trail through the Great Smoky Mountains. We did not meet Charles Darwin or Gautama Buddha, unless they were embodied by the bear who tore our campsite apart and even pulled down the backpack that was hanging from a rope 'out of reach of bears'.

John Negru has offered support and advice in the last years of this project and read two full drafts. How many publishers actually come to visit their writers? My neighbour, Patrick Doyle, has also shared many helpful discussions about dharma and human psychology. I recently saw a brown thrasher fly into a hedge on his farm. A newer friend, Janet Mason, also read the entire manuscript in 2019, and her thoughtful comments have shaped it in many ways. Along the way she reminded me that much of the story of the Trojan horse may be myth, and, still on the topic of horses, that Eeyore was not a horse but a donkey. In their daily lives these friends and neighbours exemplify the spirit of St. Francis of Assisi in a way that would probably make Gautama Buddha smile.

Appendix
How to Start Now

WELL, THIS HAS BEEN ALL VERY INTERESTING, you may think, and we have covered a lot of ground, including the origins of life, the fate of the rhinos, and even some Russian history. And, along the way, Darwin's view of natural selection and the Buddha's view on nonexistence of the ego. But, you may say, give me something practical to work with now that I am home. How might I start bringing this view into my own life? It is not possible for me to go back and meet the Buddha himself, nor do I wish to abandon my own life in a western culture.

Fair enough. It should be clear from the book that we do not need to draw a sharp distinction between secular and spiritual aspects of our lives. We can call ourselves modern people and secular people, and still live in a way that is consistent with many of the Buddha's teachings. Even he taught in many different ways, depending upon his audience, and we may reasonably assume that he would have adapted his message for modern times. We can be modern people, embrace our own cultures, and still follow the Eightfold Path. That is why I introduced you to people like Guy Bradley and Tommy Douglas, who you may be sure, do not appear as examples in any of the Buddhist teachings that I have encountered.

Here, therefore, are three places we can immediately start working with our own situation.

1. Making friends with our animal inheritance

We can begin by bringing the Eightfold Path into our life, starting perhaps with Right View. We can take this view by reminding ourselves that life is short, and it is precious because it is really all that we have. Part of Right View is to appreciate the animal nature in our human nature. We can make friends with our self, rather that fearing or repressing our innate nature. Those ape (and lobster) instincts are ancient and strong; we can be aware of them so they don't catch us by surprise. We did not choose these particular bodies or these particular behavioural tendencies or even whether or not we have a Y chromosome. This lack of choice, too, is just biological reality. Sometimes we may enjoy our bodies and our thoughts, sometimes we may not. Sometimes we may disagree with what they are suggesting and set them aside. But in any case, it is part of our challenge to see these simply for what they are, the playing-out of ancient instincts that were helpful in the distant evolutionary past. In this sense, even these ancient instincts have a kind of sacred quality to them.

2. Letting go of dwelling on the self

We can relax our intense focus on the self. This includes letting go of our personal stories about our own resentments and letting go of our temptation to dwell on the ways in which we differ from other people. Instead, can we take a more expansive view of reality. We can deliberately bring a larger view of the human condition into our lives. Life can be a journey of exploration. While on this journey, we can conduct our own lives with a certain amount of self-respect. Letting go of the self does not mean letting go of self-respect. We really can take the time each morning to eat breakfast, comb our hair, put on clean clothes, and treat ourselves with some dignity overall. And we can treat others with the same respect.

3. Immersing our mind in wild nature often

Our wild places are filled with other living beings whose futures are interwoven with ours. They are part of us, and we are part of them. So, we can intentionally immerse our mind in wild nature as often as possible. In the Buddha's day we could simply retreat to the forest. Now we must care for those very same forests so that they will continue to exist. Caring for wild places is part of the modern spiritual path. We can practice compassion not just for our close relatives, but for all living beings, particularly the ones that look very different from us. Find a wild area near you, immerse yourself often, and then become a caregiver. There are many wild areas and many wild species, and it is up to you to find the most appropriate way to serve them.

Further Reading

On Darwin, evolution, and human behaviour

General overviews

Ardrey, R. 1967. *The Territorial Imperative: A Personal Inquiry into the Animal Origins of Property and Nations*. Collins, London.

Ardrey, R. 1970. *The Social Contract: A Personal Inquiry into the Evolutionary Sources of Order and Disorder*. Atheneum, New York.

Blackmore, S. 1997. The power of the meme meme. *Skeptic* 5(2): 43-49.

Bloom, H. 1995. *The Lucifer Principle: A Scientific Expedition into the Forces of History*. The Atlantic Monthly Press, New York.

Browne, J. 1995. *Charles Darwin: Voyaging*. Princeton University Press, Princeton.

Browning, C.R. 1993. *Ordinary Men: Reserve Police Battalion 101 and the Final Solution in Poland*. Harper Perennial, New York.

Dawkins, R. 1976. *The Selfish Gene*. Oxford University Press, Oxford.

Dawkins, R. 1995. *Darwin's Dangerous Idea*. Penguin, London.

de Duve, C. 1995. *Vital Dust: Life as a Cosmic Imperative*. Basic Books, HarperCollins, New York.

Diamond, J. 1997. *Guns, Germs and Steel: The Fates of Human Societies.* W.W. Norton, New York.

Dyson-Hudson, R. and M.A. Little (eds.). 1983. *Rethinking Human Adaptation: Biological and Cultural Models.* Westview Press, Boulder, CO.

Goldsmith, T. H. 1991. *The Biological Roots of Human Nature: Forging Links Between Evolution and Behavior.* Oxford University Press, New York.

Gould, S.J. 1977. *Ever Since Darwin: Reflections in Natural History,* W.W. Norton & Company, New York.

Hardin, G. 1993. *Living Within Limits: Ecology, Economics and Population Taboos.* Oxford University Press, New York.

Ignatieff, M. 1993. *Blood and Belonging: Journeys Into the New Nationalism.* Penguin Books Canada Ltd., Toronto, ON.

Jung, C. 1971. "The Spiritual Problem of Modern Man" and "Answer to Job" reprinted as pp. 456-479 and 519-650 in Campbell, J. (ed). *The Portable Jung*, Penguin Books, UK. (translated by R.F.C. Hull)

Leakey, R. and R. Lewin. 1992. *Origins Reconsidered: In Search of What Makes Us Human.* Doubleday, New York.

Levin, H. L. 2013. *The Earth Through Time.* 10th edition. Wiley, Hoboken, NJ.

Mains, G. 1972. *The Oxygen Revolution.* David and Charles, Exeter, UK.

Margulis, L. and D. Sagan. 1986. *Microcosmos: Four Billion Years of Evolution from Our Microbial Ancestors.* Reprinted in 1997 in paperback by University of California Press, Berkeley.

Neumann, E. 1970. *The Origins and History of Consciousness.* Bollingen Series XLII. Princeton University Press, Princeton. (translated from German by R.F.C. Hull)

Oubré, A.Y. 1997. *Instinct and Revelation: Reflections on the Origins of Numinous Perception*. Gordon and Breach, Amsterdam.

Pringle, H. 1996. *In Search of Ancient North America: An Archaeological Journey to Forgotten Cultures*. Wiley and Sons, New York.

Ridley, M. 1997. *The Origins of Virtue: Human Instincts and the Evolution of Cooperation*. Viking, New York.

Staub, E. 1989. *The Roots of Evil: The Origins of Genocide and Other Group Violence*. Cambridge University Press, Cambridge.

Strahler, A.N. 1971. *The Earth Sciences*. 2nd edition. Harper and Row, New York.

Thompson, R.F. nd. *Progress in Neuroscience*. Readings from *Scientific American*. W.H. Freeman, New York.

Thompson, R.F. 1975. *Introduction to Physiological Psychology*. Harper & Row, New York.

White, T.C.R. 1993. *The Inadequate Environment*. Springer-Verlag, Berlin.

Wrangham, R. and Peterson, D. 1996. *Demonic Males: Apes and the Origins of Human Violence*. Houghton Mifflin, Boston, MA.

Specialized books and papers

Cooper, J.R., F.E. Bloom and R.H. Roth. 1978. *The Biochemical Basis of Neuropharmacology*. 3rd edition. Oxford University Press, New York.
de Duve, C. 1991. *Blueprint for a Cell: The Nature and Origin of Life*. N. Patterson, Burlington, NC.

Gould, S.J. 1977. *Ontogeny and Phylogeny*. Belknap Press of Harvard University Press, Cambridge, MA.

Knoll, A. H. 1992. The early evolution of eukaryotes: a geological perspective. *Science* 256: 622-627.

Margulis, L. 1970. *Origin of Eukaryotic Cells*. Yale University Press, New Haven, CT.

Miller, S. L. 1953. A production of amino acids under possible primitive earth conditions. *Science* 117: 528-529.

Miller, S. L. 1955. Production of some organic compounds under possible primitive Earth conditions. *Journal of the American Chemical Society* 77: 2351-2361.

Moacanin, R. 1986. *Jung's Psychology and Tibetan Buddhism: Western and Eastern Paths to the Heart*. Wisdom Publications, London.

Morowitz, H. J. 1968. *Energy Flow in Biology: Biological Organization as a Problem in Thermal Physics*. Academic Press, New York.

Okihana, H. and C. Ponnamperuma. 1982. A protective function of the coacervates against UV light on the primitive Earth. *Nature* 299: 347-349.

Orgel, L.E. 1973. *The Origins of Life: Molecules and Natural Selection*. Wiley and Sons, New York.

Robinson, J. M. 1990. Lignin, land plants, and fungi: biological evolution affecting Phanerozoic oxygen balance. *Geology* 15: 607-610.

Schopf, J. W. and E. S. Barghoorn. 1967. Alga-like fossils from the early Precambrian of South Africa. *Science* 156: 508-512.

Taylor, T.N. 1988. The origin of land plants. Some answers, more questions. *Taxon* 37: 805-833.

Taylor, T. N. 1990. Fungus associations in the terrestrial paleoecosystem. *Trends in Ecology and Evolution* 5: 21-25.

On Buddhism and consciousness

Batchelor, S. 1997. *Buddhism Without Beliefs: A Contemporary Guide to Awakening*. Berkley Publishing Group, New York.

Further Reading

Chodron, P. 1997. *When Things Fall Apart: Heart Advice for Difficult Times*. Shambhala Publications, Boston, MA.

Dalai Lama. 1985. *Opening the Eye of New Awareness*. Wisdom Publications, London. (translated by D.S. Lopez, Jr. with Jeffrey Hopkins)

Dogen, Z. and K. Uchiyama. 1983. *Refining Your Life: From the Zen Kitchen to Enlightenment*. Weatherhill, New York.

Fields, R. 1992. *How the Swans Came to the Lake: A Narrative History of Buddhism in America*. 3rd edition. Shambhala Publications, Boston, MA.

Gampopa. *The Jewel Ornament of Liberation*. 1986. Shambhala Publications, Boston, MA. (translated and annotated by H.V. Guenther)

Khyentse, D. 1993. *Enlightened Courage: An Explanation of Atisha's Seven Point Mind Training*. Snow Lion Publications, Ithaca, NY. (translated by Padmakara Translation Group)

Kongtrul, J. 1987. *The Great Path of Awakening: A Commentary on the Mahayana Teaching of the Seven Points of Mind Training*. Shambhala Publications, Boston, MA. (translated by K. McLeod)

Kongtrul, J. 1992. *Cloudless Sky: The Mahamudra Path of the Tibetan Buddhist Kagyu School*. Shambhala Publications, Boston, MA. (translated by T. Drasczyk and A. Drasczyk)

Nalanda Translation Committee. 1995. *The Life of Marpa the Translator*. Shambhala Publications, Boston, MA.

Sogyal Rinpoche. 1992. *The Tibetan Book of Living and Dying*. (P. Gaffney and A. Harvey eds.) Harper, San Francisco.

Suzuki, S. 1970. *Zen Mind, Beginner's Mind*. Weatherhill, New York.

Trungpa, C. 1985. *Born in Tibet*. 3rd edition. Shambhala Publications, Boston, MA.

Trungpa, C. 1985. *Journey Without Goal: The Tantric Wisdom of the Buddha*. Shambhala Publications, Boston, MA.

Trungpa, C. 1991. *The Heart of the Buddha*. Shambhala Publications, Boston, MA.

Trungpa, C. 1992. *Transcending Madness: The Experience of the Six Bardos*. Shambhala Publications, Boston, MA.

Tulku, T. 1984. *Knowledge of Freedom: Time to Change*. Dharma Press, Oakland, CA.

Other books referred to in text or recommended by section

The Men and Their Theories

Desmond, A. and J. Moore. 1991. *Darwin*. Warner Books, New York.

Schumann, H.W. 2004. *The Historical Buddha*. Motilal Banarsidass Publishers, Delhi. (translated from the German by M. O'C. Walsh)

1. Craving for Resources: Desire, Dissatisfaction, and Suffering

Brackenridge, J.B. and R.M. Rosenberg. 1970. *The Principles of Physics and Chemistry*. McGraw-Hill Book Company, New York.

Ferris, T. 1989. *Coming of Age in the Milky Way*. Doubleday, New York.

Marschner, H. 1995. *Mineral Nutrition of Higher Plants*. 2nd edition. Academic Press, London.

Morowitz, H. J. 1968. *Energy Flow in Biology*. Academic Press, New York.

Orians, G.H. 1998. Human behavioral ecology: 140 years without Darwin is too long. *Bulletin of the Ecological Society of America* 79(1): 15-28.

Smith, A. 1776. *An Enquiry into the Nature of the Wealth of Nations*. in M.J. Adler (ed.). 1990. *Great Books of the Western World*, Vol. 36.

Encyclopaedia Britannica, Chicago, IL.

Vitousek, P.M., J. Aber, R.W. Howarth, G.E. Likens, P.A. Matson, D.W. Schindler, W.H. Schlesinger and G.D. Tilman. 1997. Human alteration of the global nitrogen cycle: causes and consequences. *Issues in Ecology* 1 (Spring).

2. Living an Illusion: Mind as Cocoon

Cairns-Smith, A.G. 1985. *Seven Clues to the Origin of Life*. Cambridge University Press, Cambridge, UK.

Day, W. 1984. *Genesis on Planet Earth*. 2nd edition. Yale University Press, New Haven, CT.

Margulis, L. and M. F. Dolan. 1997. Swimming against the current. *The Sciences* (January/February): 20-25.

Matthews, B. 1983. *Craving and Salvation: A Study in Buddhist Soteriology*. Wilfred Laurier University Press, Waterloo, ON.

Miller, S. L. 1974. The first laboratory synthesis of organic compounds under primitive Earth conditions. pp. 228-242 in J. Neyman (ed.), *The Heritage of Copernicus: Theories "Pleasing to the Mind"*. MIT Press, Cambridge, MA.

Ray, R. 2008. *Touching Enlightenment: Finding Realization in the Body*. Sounds True, Boulder, CO.

Trivers, R.L. 1976. Foreward to Dawkins, R. 1976. *The Selfish Gene*. Oxford University Press, Oxford. (Granada edition 1978)

Trungpa, C. 1987. *Glimpses of Abhidharma*. Shambhala Publications, Boston, MA.

3. The Primate Prison: The Origin of Self

Sacks, O. 1990. *The Man Who Mistook His Wife for a Hat and Other Clinical Tales*. Harper Perennial, New York.

Trungpa, C. 1972. *Karma Seminar.* Transcript of a seminar given at Karmê-Chöling, September 1972. Vajradhatu Publications, Boulder, CO.

4. Selective Memory: Maintaining the Illusion

Bodhi, B. 1994. *The Noble Eightfold Path.* Paravati Publications, Onalaska, WA.

Dye, T.R. and H. Zeigler. 1987. *The Irony of Democracy.* 7th edition. Brooks/Cole, Monterery, CA.

Frazer, Sir. J. G. 1922. *Selections from the Golden Bough: A Study in Magic and Religion.* pp. 1-78 in M.J. Adler (ed.). 1990. *Great Books of the Western World*, Vol. 58. Encyclopaedia Britannica, Chicago, IL.

Ginsberg, A. 1994. *Cosmopolitan Greetings. Poems 1986-1992.* HarperCollins, New York.

Hall, J.A. 1983. *Jungian Dream Interpretation: A Handbook of Theory and Practice.* Inner City Books, Toronto, ON.

Heffernan, M. 2011. *Willful Blindness: Why We Ignore the Obvious at Our Peril.* Doubleday Canada, Toronto, ON.

Howe, M.L. and L.M. Knott. 2015. The fallibility of memory in judicial processes: Lessons from the past and their modern consequences. *Memory* 23 (5): 633-656.

Shalamov, V. 1980. *Kolyma Tales.* W.W. Norton, New York. (translated from Russian by J. Glad; quotations from paperback edition 1982)

5. The Urge to Impress: Priests, Kings, and Dominance Hierarchies

Baker, R. 1996. *Sperm Wars: The Science of Sex.* HarperCollins, Toronto, ON.

Boone, J.L. 1983. Noble family and expansionist warfare in the late middle ages: A sociological approach. pp. 79-96 in R. Dyson-Hudson and M.A. Little (eds.) *Rethinking Human Adaptation and Cultural Models*. Westview Press, Boulder, CO.

Brown, J.L. 1975. *The Evolution of Behaviour*. Norton and Co., New York.

Cross, T.P. and C.H. Slover (eds.). 1936. *Ancient Irish Tales*. Henry Holt and Co. Reprinted in 1996 by Barnes and Noble, New York.

Dewsbury, D.A. 1982. Dominance rank, copulatory behaviour, and differential reproduction. *The Quarterly Review of Biology* 57: 135-159.

Diamond, J. 1992. *The Third Chimpanzee*. Harper, New York.

D'Souza, D. 1995. *The End of Racism: Principles for a Multicultural Society*. The Free Press, New York.

Durant, W. 1944. *The Story of Civilization: Part III Caesar and Christ*. Simon and Schuster, New York.

Gibbon, E. 1776. *The Decline and Fall of the Roman Empire*. in M.J. Adler (ed.). 1990. *Great Books of the Western World*, Vol. 37. Encyclopaedia Britannica, Chicago, IL.

Hayden, B. and J. Spafford. 1993. The Keatley Creek site and corporate group archeology. *BC Studies* 99 (Autumn): 106-139.

Herrnstein, R. and Murray, C. 1994. *The Bell Curve: Intelligence and Class Structure in American Life*. Free Press, New York.

Hayden, B. and J. M. Ryder. 1991. Prehistoric cultural collapse in the Lillooet Area. *American Antiquity* 56: 50-65.

Keddy, A.C., R.M. Seyfarth, and M.J. Raleigh. 1989. Male parental care, female choice, and the effect of an audience in vervet monkeys. *Animal Behaviour* 38: 262-271.

Marra, P.P. and R.T. Holmes. 1997. Avian removal experiments: do they test for habitat saturation or female availability? *Ecology* 78: 947-952.
Miele, F. 1996. The immortal animal. *Skeptic* 4: 42-49.

Morris, D. 1967. *The Naked Ape*. McGraw Hill, New York.

Peterson, J. B. 2018. *12 Rules for Life: An Antidote to Chaos*. Random House Canada, Toronto, ON.

Plutarch. nd. *The Lives of the Noble Grecians and Romans* (Dryden translation). in M.J. Adler (ed.). 1990. *Great Books of the Western World*, Vol. 13. Encyclopaedia Britannica, Chicago, IL.

Rushton, J.P. 1995. *Race, Evolution and Behavior*. Transaction, New Brunswick, NJ.

Small, M. F. 1993. *Female Choices: Sexual Behavior of Female Primates*. Cornell University Press, Ithaca, NY.

Stewart, R.E. and J.W. Aldrich. 1951. Removal and re-population of breeding birds in a spruce-fir forest community. *Auk* 68: 471-482.

Struhsaker, T.T. 1967a. Social structure among vervet monkeys (*Ceropithecus aethiops*). *Behaviour* 29: 83-121.

Struhsaker, T.T. 1967b. Ecology of vervet monkeys (*Ceropithecus aethiops*) in the Masai-Amboseli game reserve, Kenya. *Ecology* 48: 891-904.

6. Killing Minds and Killing Fields: Interference, Competition, and Agression

Bostom, A.G. (ed.). 2005. *The Legacy of Jihad: Islamic Holy War and the Fate of Non-Muslims*. Prometheus Books, Amherst, NY.

Canny, M.J. 1998. Transporting water plants. *American Scientist* 86: 152-159.

Frankland, N. 1970. *Bomber Offensive: the Devastation of Europe*. Ballantine Books, New York.

Freud, S. 1915. *The Major Works of Sigmund Freud*. in M.J. Alder (ed.). 1990. *Great Books of the Western World*, Vol. 54. Encyclopaedia Britannica, Chicago, IL.

Handford, S.A. 1951. *Caesar: The Conquest of Gaul.* Penguin Books, Harmondsworth, UK.

Ignatieff, M. 1993. *Blood and Belonging.* Penguin Books Canada Ltd., Toronto, ON.

Keegan, J. 1989. *The Second World War.* Viking Penguin, New York. (Amsterdam edition published 1990)

King, J. 1997. *Reaching for the Sun: How Plants Work.* Cambridge University Press, Cambridge.

Moorehead, A. 1945. *Eclipse.* Hamish Hamilton Ltd., London. (Republished in new edition 1967)

Nicholas, L.H. 1994. *The Rape of Europa: The Fate of Europe's Treasures in the Third Reich and the Second World War.* Knopf, New York.

Saul, J.R. 1995. *The Unconscious Civilisation.* House of Anansi Press, Concord, ON.

Speer, A. 1970. *Inside the Third Reich.* Avon Books, New York. (translated from the 1969 German edition edited by R. and C. Winston)

Zamoyski, A. 2004. *Moscow 1812: Napoleon's Fatal March.* HarperCollins, New York.

7. Insatiable Consumption: When Big Brains Meet Big Animals

Armstrong, A.O. 1966. *St. Francis of Assisi: A Concise Biography.* American R.D.M. Corporation, New York.

Binford, M.W., M. Brenner, T.J. Whitmore, A. Higuera-Gundy, E.S. Deeveyand B. Leyden. 1987. Ecosystems, paleoecology and human disturbance in subtropical and tropical America. *Quaternary Scientific Review* 6: 115-128.

Chaber A-L, S. Allebone-Webb, Y. Lignereux, A.A. Cunningham and J.M. Rowcliffe. 2010. The scale of illegal meat importation from Africa to Europe via Paris. *Conservation Letters* 3: 317-321.

Diamond, J. 1994. Ecological collapses of past civilizations. *Proceedings of the American Philosophical Society* 138: 363-370.

Diamond, J. 1995. Easter's End. *Discover* (August): 62-69.

Ehrlich, A. and P. Ehrlich. 1981. *Extinction: The Causes and Consequences of the Disappearance of Species*. Random House, New York.

Firestone, R.B. (and 25 others). 2007. Evidence for an extraterrestrial impact 12,900 years ago that contributed to the megafaunal extinctions and the Younger Dryas cooling. *Proceedings of the National Academy of Sciences of the United States of America* 41: 16016-16021.

Harrison, R.D., S. Tan, J.B. Plotkin, F. Slik, M.Detto, T. Brenes, A. Itoh and S.J. Davies. 2013. Consequences of defaunation for a tropical tree community. *Ecology Letters* 16: 687-694.

Hornaday, W.T. 1913. *Our Vanishing Wild Life: Its Extermination and Preservation*. New York Zoological Society, New York.

Hughes, J.D. and J.V. Thirgood. 1982. Deforestation, erosion, and forest management in ancient Greece and Rome. *Journal of Forest History* 26: 60-75.

Mann, C.C. 2005. *1491: New Revelations of the Americas before Columbus*. Knopf, New York.

Martin, P.S., and R.J. Klein. 1984. *Quaternary Extinctions: A Prehistoric Revolution*. The University of Arizona Press, Tucson, AZ.

McIver, S.B. 2003. *Death in the Everglades: The Murder of Guy Bradley, America's First Martyr to Environmentalism*. Florida History and Culture Series. University Press of Florida, Gainesville, FL.

Turner, F. 1994. *Beyond Geography: The Western Spirit Against the Wilderness*. Rutgers University Press, NJ.

Vitousek, P.M., P.R. Ehrlich, A.H. Ehrlich and P.A. Matson. 1986. Human appropriation of the products of photosynthesis. *Bioscience* 36: 368-373.

8. Getting Along: An Ecological View of Compassion

Fogelman, E. 1994. *Conscience and Courage: Rescuers of Jews During the Holocaust*. Anchor Books, Doubleday, New York.

Gyatso, G.K. 1991. *The Bodhisattva Vow*. Tharpa Publications, London.

Lam, V. 2011. *Tommy Douglas*. Penguin Canada, Toronto, ON.

Leopold, A. 1949. *A Sand County Almanac*. Oxford University Press, London.

Merton, T. 1967. *Mystics and Zen Masters*. Dell Publishing, New York.

Miele, F. 1996. The immortal animal. *Skeptic* 4: 42-49.

Thich Nhat Hanh. 1988a. *The Heart of Understanding*. Parallax Press, Berkeley, CA.

Thich Nhat Hanh. 1998b. Resting in the river. *Shambhala Sun* (March): 42-51.

9. A Brief History of Life: Co-operation and Community

Cairns-Smith, A.G. 1985. *Seven Clues to the Origin of Life*. Cambridge University Press, Cambridge, UK.
Dawkins, R. 1998. The Selfish Cooperator. pp. 38-47 in *Commemorative Lecture, International Cosmos Prize* 1997. The Commemorative Foundation of the International Garden and Greenery Exposition, Osaka, Japan, 1990.

Keddy, P.A. 1990. Is mutualism really irrelevant to ecology? *Bulletin of the Ecological Society of America* 71: 101-102.

Keddy, P.A. 2010. *Wetland Ecology: Principles and Conservation* (2nd edition). Cambridge University Press, Cambridge, UK.

Keddy, P.A. 2017. *Plant Ecology: Origins, Processes, Consequences*. Cambridge University Press, Cambridge, UK.

Lane, N. 2010. *Life Ascending: Ten Great Inventions of Evolution*. W.W. Norton, New York.

Lane, N. 2015. *The Vital Question: Energy, Evolution and the Origins of Complex Life*. W.W. Norton, New York.

Leopold, A. 1949. *A Sand County Almanac*. Oxford University Press, Oxford.

McHarg, I. L. 1969. *Design with Nature*. Natural History Press for American Museum of Natural History, Garden City, NJ.

Sender, R., S. Fuchs, and R. Milo. 2016. Revised estimates for the number of human and bacteria cells in the body. PLoS Biology 14(8): e1002533.

10. Meditation in Action: Seeing Through the Simulation

Kumar, S.M. 2009. *The Mindful Path Through Worry and Rumination: Letting Go of Anxious and Depressive Thoughts*. New Harbinger Publications, Oakland, CA.

Lama Surya Das (ed.). 1993. *Natural Great Perfection: Vajra Songs and Dzogchen Teachings*. Dzogchen Foundation of America, Bloomfield, NY. Merton, T. 1975. *The Asian Journal of Thomas Merton*. New Directions Publishing Corp., New York. Edited from his original notebooks by N. Burton, P. Hart and J. Laughlin

Nyoshul Khenpo Rinpoche. 1993. *Natural Great Perfection*. Dzogchen Foundation of America, Bloomfield, NY. (edited by Lama Surya Das) Parchelo, R. 2012. *Walk Like a Mountain: The Handbook of Buddhist Walking Practice*. Sumeru, Ottawa.

Thich Nhat Hanh. 1988. *The Heart of Understanding*. Parallax Press, Berkeley, CA.

Trungpa, C. 1985. *Journey Without Goal: The Tantric Wisdom of the Buddha*. Shambhala Publications, Boston, MA.

Trungpa, C. 1991. *The Heart of the Buddha*. Shambhala Publications, Boston, MA.

11. Enlightened Society: The Evolutionary Imperative

Bhikkhu Bodhi (ed.). 2016. *The Buddha's Teachings on Social and Communal Harmony*. Wisdom Publications, Somerville, MA.

Blake, W. pp. 40-111 in M.A. Abrams (ed.). 1974. *The Norton Anthology of English Literature*. 3rd edition. W. W. Norton, New York.

David-Neel, A. and Lama Yogden. 1987. *The Superhuman Life of Gesar of Ling*. Shambhala Publications, Boston, MA.

Hayward, J. 1995. *Sacred World: A Guide to Shambhala Warriorship in Daily Life*. Bantam Books, New York.

Homer. *The Odyssey*. 1992. Wordsworth Classics, Ware, UK. (translated by T.E. Shaw)

Jung, C. 1971. The spiritual problem of modern man. Reprinted as pp. 456-479 in J. Campbell (ed.) *The Portable Jung*, Penguin Books, UK. (translated by R.F.C. Hull)

Manual, F.E. (ed.). 1966. *Utopias and Utopian Thought*. Beacon Press, Boston, MA.

Matthews, J. 1989. *The Elements of the Arthurian Tradition*. Element Books Ltd., Shaftesbury, Dorset, UK.

Trungpa, C. 1985. *Born in Tibet*. 3rd edition. Shambhala, Boston, MA.

Figure Credits

Page 26 by Ian Keddy

Page 28 from Taube, M. 1982. *Evolution of Matter and Energy in Cosmic and Planetary Scale.* Killwanger, Swtizerland (self-published) in T. Ferris. 1989. *Coming of Age in the Milky Way.* Doubleday, New York.

Page 36 by Cathy Keddy

Page 80 by Cathy Keddy

Page 90 adapted from illustration by S. Biesty. 1993. *Stephen Biesty's Cross-Sections: Man-of-War.* Scholastic Canada Ltd., Richmond Hill.

Page 94 White-footed racket tail, p. 388 in C. Darwin. 1871. *The Descent of Man, and Selection in Relation to Sex.* John Murray, London.

Page 116 Giant ground sloth by J. Smit in H. N. Hutchinson. 1892. *Extinct Monsters: A Popular Account of Some of the Larger Forms of Ancient Animal Life.* Chapman & Hall, London.

Page 124 Bicknell's thrush by Adelaide Tyrol

Page 186 Wood frog by Howard Coneybeare

Page 208 *Magnolia macrophylla* (bigleaf magnolia), Figure 315 in C. S. Sargent. 1922. *Manual of the Trees of North America.* Houghton Mifflin, New York.

Index

Algonquin Park   142, 211
Africa   29, 37, 60, 79, 106-107, 117, 118, 120, 121, 126, 147, 165, 191, 193
Appalachian Mountains (*see* Great Smoky Mountains)
Appalachian Trail   5, 165, 182, 185, 192, 211

barnacles   16, 41, 82-83, 95, 115,
Batchelor, Stephen   13, 25
Bertolt Brecht   106
Bhikkhu Bodhi   9, 24, 65, 169, 206
Bible   133, 135, 139-140, 146
Bicknell's thrush   42, 127, 128, 182, 212
birds   41, 42, 115, 123-124, 127, 128, 145, 169, 182, 187, 189, 193, 208, 209, 211, 212
Blake, William   51, 57, 192, 196-198
Blanding's turtles   127, 156, 189
Bradley, Guy   123-124, 189, 215
Buddha (*see* Siddhartha Gautama)
Buddha nature   174-175, 188-190, 207, 209

Camelot   192, 197
carbon   30, 149-151, 158-160, 162
cells   17, 30, 133, 138, 145, 148-157, 162-163, 193,
cocoon   39-50, 65, 76, 180, 195, 202
compassion (*see also* idiot compassion)   6, 9, 35, 125, 130-131, 133-144, 161, 168-169, 178-179, 185, 193, 196, 198, 202, 216
competition   79-82, 99-113, 135-136, 157

complaining   40, 146, 161, 167, 173-174
co-operation
    among selfish genes   133-135
    biological basis   136-137, 146-148
    in eukaryotic cells   152-154, 193
    in multicellular organisms   154-156
    in relation to compassion   136-137
    in relation to social organization   138-139, 143-144, 145, 157-158, 162
    when hiking   145
craving   22, 25-38, 39-42, 49, 77, 89, 100, 117, 189, 191, 192, 209

Darwin, Charles
    birth   15
    books
        Descent of Man, The   46, 95
        Insectivorous Plants   32
        monograph on barnacles   82-83
        Origin of Species, The   16, 115
        Voyage of the Beagle, The   115
    hierarchies   91, 199, 206
    human behavior   5-7, 37, 40
    human suffering   6, 170-171, 196
    life incidents   15-16, 91, 159, 170
    origins of mind   40-41, 45-50, 53-54, 172, 180
    natural selection   17-21, 29-30, 40-46, 48, 76, 179
    representing modern biology   7, 16-17, 29, 138, 153, 157, 193, 196, 207
    sexual selection   94, 95
    social Darwinism   138, 170-171
    survival of the fittest   146-147 (*see also* natural selection)
    warfare   109, 206
defaunation   115-121
Dawkins, Richard   20
Descartes   51
Descent of Man, The   46, 94, 95
Douglas, Tommy   143-144, 189, 215
dukkha   21, 25-38, 49, 115, 147, 192, 194

ego (*see also* illusion)
    ape ego (the self)   51-63, 74-76, 112, 168-170, 177-178, 180, 190, 198
    complaining   40, 173
    distraction   69
egotism in bird watching   187
    egotism in hiking   9, 40
    egotism in science   8
    egotism in Buddhist practices   187
    emptiness of   43, 52, 140, 190, 215
    in depth psychology   61-62
    role of memory   65-76
Eightfold Path   5, 12, 24, 65, 124, 140, 145, 166-167, 169, 171, 192, 197, 205, 215
enlightenment   15, 52, 53, 91, 138, 143, 163, 168, 175, 180, 186, 188, 190, 198, 205, 209
Epstein, Mark   66
eukaryotic cells   148, 152-154, 157, 193

female choice   92-97
French Revolution   13, 76, 89, 194
Freud, Sigmund   99
frogs   9, 11, 21, 123, 125, 142, 175, 186

Ganges River   27, 66, 125, 159, 189
Genghis Khan   86-87, 202
Great Smoky Mountains   5, 42, 129, 141, 160, 192, 209, 212
gulag   65, 67, 68-71, 74-76, 194

Heart Sutra   43
Heffernan, Margaret   74, 158
hierarchies   7, 77-97, 199-203
hiking   9, 25, 27, 38, 40, 49, 77, 145, 160, 162-162, 165, 166, 174-175, 183-184, 188, 192, 198-199, 207-208
Hinayana   168-176
hooded warbler   42, 192, 208, 209
homosexuality, natural selection   134-135
hydrothermal vents   151

icons   106, 183
idiot compassion   140-142, 147, 201
Ignatieff, Michael   110
illusion
    analogy of cocoon   39-49, 65, 76, 180, 195, 202
    as sense of self   43, 51-54, 74-75, 161-162
    as propaganda   71-72
    as simulation of reality   43, 51-61, 172-174, 180, 189-190
    as source of suffering   23, 55
    Descartes   51
    desperation   55-57
    evolutionary origins   40-41, 43, 48-49, 168
    fear   43, 46-49
    Heart Sutra   43
    in hiking   40, 62, 187
    in meditation   23, 54-56, 58-59, 111-112, 165, 168, 173-174, 198
    inaccuracy of   45-46, 56-46, 47-49
    memory   65-72
    Shakespeare   52, 56-57
    skandhas   42
interbeing   117, 158-163, 174, 193, 216
Islam   107-109, 135-136, 183

Jesus   9, 10, 49, 55, 58, 89, 133, 139, 143, 160, 182, 187
jihad   107-109, 135-136
Job   139-140, 146, 150-151
Jung, Carl   58, 61, 77, 180, 184

kindness (*see* compassion)
Kumar, Sameet   175

large animals   115-121
looting   104-109, 183, 195
lobsters   88, 91, 95, 156, 177

Medina   135-136
Mahayana   168, 169, 176-179
Mao   89, 206

Maori   120
matriarchy   87
McHarg, Ian   160-162
meditation   5-6, 8, 10, 12, 23-24, 42, 44, 52, 55-59, 66, 75, 111-112, 125-127, 140, 165-168, 171-176, 181-184, 188, 194, 196-198, 207-209
megafauna   115-121
membranes   29-30, 33, 151-152
memory   43, 56, 62, 65-76, 200
Merton, Thomas   169-170
Miller/Urey experiment   150

nitrogen   30-33, 104, 149-150, 159
Napoleon   76, 89, 90, 106
Naropa   66-67, 188
Nova Scotia   126-127, 132, 160, 211
Nyoshul Khenpo   190

overpopulation   35-38, 125, 147, 156, 201
oxygen   29-32, 103, 149, 155, 161-162

Pali   15, 42, 203, 206
patriarchy   86, 92 (*see also* hierarchies)
peacocks   97
pillage (*see* looting)
Pollyanna   72 (*see also* wilful blindness)
Priam (King of Troy)   199-203

Ray, Reggie   44-45
Right livelihood   24, 166
rigpa   190
Roman Empire   79, 83-85, 100
Russian Revolution   75, 89

samsara
    causes   12, 21-22, 25-38, 170-171
    escape   22-24, 173-175, 189-190
    in relation to conflict and competition   99-113
    in relation to hierarchies   77-97

    in relation to interbeing   161-162
    in science   8
    wishful thinking   147
Sanskrit   25, 26, 42, 172, 174
Schumann, H.W. (*including his book,* The Historical Buddha)   15, 91, 124, 166, 203-205
Second World War   69-70, 74, 101-102, 104-106
serotonin   41, 88-89, 156
sexual misconduct   73, 82, 84-85, 93-97, 189, 205
sexual selection   81-83, 93-97
selfish genes   79, 110, 133-137, 168, 170-171, 178-179, 193
Shakespeare   33, 34, 35, 52, 54, 56-57, 68, 75, 173
Shalamov, Varlam   65, 68-71
Shambhala   192, 197, 207
Shantideva   115, 124-125, 127, 185
Siddhartha Gautama (*also when referred to as* the Buddha)
    birth   15
    death   159, 175
    dukkha   21-22, 25, 27
    Eightfold Path   24, 166
    hiking trip   5, 166, 207
    human behaviour   6, 8, 12, 35, 49, 196
    hungry ghosts   35
    governance   203-204, 206-207
    illusion   39-44, 51-58
    karma   13
    life incidents   15, 27, 49, 66, 76, 91, 160
    meditation   23-24, 111-112, 165
    rules for sangha   124, 113, 124-188, 167, 205
    samsara   8, 12, 21-22, 173
    self-examination   111, 171
    successor   204
    teaching   8, 12, 21-24, 51, 75, 140, 145, 180-182, 194
    transmission of teachings after death   108-109, 189, 203-205
    walking   174-175
    wild nature   9-10, 97, 115-116, 124-125
simulation (*see* illusion)
Snyder, Gary   162
speciesism   127

Stalin   66, 68, 71, 92, 194, 206
status   34, 89, 96, 138, 173, 187 (*see* hierarchies)
Suzuki Roshi   56, 165, 175, 181, 186, 188, 189,
St. Francis of Assisi   9, 181, 183, 184, 213

Tantra   190 (*see* Vajrayana)
ten positive actions   130-131
three poisons   8
Tibet   12, 91, 109, 177, 183, 184, 187, 188, 197-198, 204
Tilopa   66, 67, 189
tonglen   177-179
Troy   200-202
Trungpa, Chögyam
    ape instinct   165
    camping   187, 212
    cocoon   44
    Great Stupa of Dharmakaya   188
    life incidents   184, 188, 197-198, 204
    portrait   182
    sacred path of warriorship   12, 44
    Shambhala   197
    teaching   75, 165, 168, 169, 172, 206
Tuchman, Barbara   109, 199, 200, 201, 203
tyrants   33, 80, 83, 84, 87, 142, 199-203 (*see also* hierarchies)

Vajrayana   180-184
visualization   139, 180-184, 188

warfare   49, 66, 99-106, 107-111, 202
wilderness   7, 9, 35-38, 126-127, 130-131, 132, 188, 197, 206
wilful blindness   12, 74, 147

xenophobia   136

Y chromosome   7, 86-87, 205, 216

www.ingramcontent.com/pod-product-compliance
Lightning Source LLC
Chambersburg PA
CBHW032022230426
43671CB00005B/173